QD
257.7
.B394
1998

D1082918

BELL LIBRARY - TAMU-CC

The Beilstein System
Strategies for Effective Searching

Stephen R. Heller, Editor

U.S. Department of Agriculture

American Chemical Society, Washington, DC

Library of Congress Cataloging-in-Publication Data

The Beilstein system: strategies for effective searching / edited by Stephen R.
Heller.
   p. cm.
Includes index.
ISBN 0-8412-3523-6 (alk. paper)

   1. Beilstein, Friedrich Konrad, 1838-1906. Beilsteins Handbuch der
organischen Chemie. 2. Chemistry, Organic--Databases.
3. Information storage and retrieval systems--Chemistry.
4. Database searching. I. Heller, Stephen R., 1943-   .
QD257.7.B394   1997
025.06'547--dc21   97-35536
                                                           CIP

The paper used in this publication meets the minimum requirements of American
National Standard for Information Sciences—Permanence of Paper for Printed
Library Materials, ANSI Z39.48-1984.

Copyright © 1998 American Chemical Society

All Rights Reserved. Reprographic copying beyond that permitted by Sections
107 or 108 of the U.S. Copyright Act is allowed for internal use only, provided
that the per-chapter fee of $20.00 base + $.25/page is paid to the Copyright
Clearance Center, Inc., 222 Rosewood Drive, Danvers, MA 01923, USA. Republi-
cation or reproduction for sale of pages in this book is permitted only under license
from ACS. Direct these and other permission requests to ACS Copyright Office,
Publications Division, 1155 16th St., N.W., Washington, DC 20036.

The citation of trade names and/or names of manufacturers in this publication is
not to be construed as an endorsement or as approval by ACS of the commercial
products or services referenced herein; nor should the mere reference herein to
any drawing, specification, chemical process, or other data be regarded as a license
or as a conveyance of any right or permission to the holder, reader, or any other
person or corporation, to manufacture, reproduce, use, or sell any patented inven-
tion or copyrighted work that may in any way be related thereto. Registered
names, trademarks, etc., used in this publication, even without specific indication
thereof, are not to be considered unprotected by law.

PRINTED IN THE UNITED STATES OF AMERICA

This book is dedicated to Joshua, Adam, and Matthew—three lovely boys—a doctor, a lawyer, and a Webmaster.

Advisory Board

Mary E. Castellion
ChemEdit Company

Arthur B. Ellis
*University of Wisconsin
at Madison*

Jeffrey S. Gaffney
Argonne National Laboratory

Gunda I. Georg
University of Kansas

Lawrence P. Klemann
Nabisco Foods Group

Richard N. Loeppky
University of Missouri

Cynthia A. Maryanoff
*R. W. Johnson Pharmaceutical
Research Institute*

Roger A. Minear
*University of Illinois
at Urbana–Champaign*

Omkaram Nalamasu
AT&T Bell Laboratories

Kinam Park
Purdue University

Katherine R. Porter
Duke University

Douglas A. Smith
The DAS Group

Martin R. Tant
Eastman Chemical Company

Michael D. Taylor
*Parke-Davis Pharmaceutical
Research*

Leroy B. Townsend
University of Michigan

William C. Walker
DuPont Company

About the Editor

STEPHEN R. HELLER is a Research Scientist in the U.S. Agricultural Research Service's Beltsville Laboratory, where, since 1985, he has been the Informatics Project Leader for the Plant Genome Research Program. He directs this worldwide genome-mapping database and retrieval system for agronomically important crops, and also develops databases and computer systems. He is Founder and Chairman of the annual Plant and Animal Genome Conference.

From 1973 to 1985, Dr. Heller was a Project Manager in the Environmental Protection Agency (EPA), where he managed the design, development, and implementation of the Chemical Information System (CIS), a joint project of the EPA and the National Institutes of Health which had thousands of users worldwide. He also managed the design, development, and implementation of the Mass Spectral Database, which, like the CIS, has proven to be a commercial success. In addition, he has developed and coordinated several other scientific databases, modeling systems, and expert system programs.

Dr. Heller received a Ph.D. in physical organic chemistry from Georgetown University, Washington, DC, in 1967, and a B.S. in chemistry from the State University of New York, Stony Brook, in 1963.

Dr. Heller is an international authority on scientific numeric and factual databases, electronic publishing, and chemical information. During the past 30 years, he has published over 145 papers in leading scientific journals and has served as the editor of five chemistry and spectral data books. He has been a consultant to chemical, chemical information, database, and publishing companies. He is the past chairman of the International Union of Pure and Applied Chemistry (IUPAC) Committee on Chemical Databases and is a member of the IUPAC Committee on Printed and Electronic Publications.

Dr. Heller has served as a member of the ACS Software Advisory Board (1988–1996), and as a Councilor of the ACS Division of Computers in Chemistry (1988–1995). He is the Software Review Editor of the *Journal of Chemical Information and Computer Sciences* and the editor of the *Trends in Analytical Chemistry* Internet column.

Dr. Heller has received numerous awards and honors, including scholarships and fellowships from the National Science Foundation, National Academy of Sciences, National Institutes of Health, and the North Atlantic Treaty Organization. He was awarded the EPA Gold Medal in 1976. He is a member of the ACS, American Society for Mass Spectrometry, Institute of Electrical and Electronics Engineers, American Association for the Advancement of Science, Society for Applied Spectroscopy, and Sigma Xi.

Contents

Contributors

John M. Barnard
    barnard@bci1.demon.co.uk
    Barnard Chemical Information Ltd., 46 Uppergate Road, Stannington,
    Sheffield S6 6BX, U.K.

Andreas Barth
    ab@fiz-karlsruhe.de
    STN International, FIZ Karlsruhe, D–76344,
    Eggenstein-Leopoldshafen, Germany

Roger Beckman
    beckmanr@cluster.ucs.indiana.edu
    Chemistry Library, Indiana University, Bloomington, IN 47405

Stephen R. Heller
    srheller@gig.usda.gov
    Agricultural Research Service, U.S. Department of Agriculture,
    Beltsville, MD 20705–2350

Alexander J. Lawson
    alawson@beilstein.com
    Beilstein Information Systems GmbH, Varrentrappstrasse 40-42,
    D–60486 , Frankfurt-am-Main, Frankfurt, Germany

Reiner Luckenbach
rxl22@xtrn.org
Bahnhofstrasse 22, D–61476, Kronberg, Germany

Ken Rouse
krouse@vms.macc.wisc.edu
Chemistry Library, University of Wisconsin, Madison, WI 53706

Dirk Walkowiak
dwalkowiak@beilstein.com
Softron GmbH, Dornierstrasse 4, D–82110 , Germering, Germany

Wendy Warr
wendy@warr.com
Wendy Warr & Associates, 6 Berwick Court, Holmes Chapel,
Cheshire CW4 7HZ, England

Janusz L. Wisniewski
jwisniewski@beilstein.com
Beilstein Information Systems GmbH, Varrentrappstrasse 40-42,
D–60486 , Frankfurt-am-Main, Frankfurt, Germany

Bernd Wollny
bwollny@beilstein.com
Beilstein Information Systems GmbH, Varrentrappstrasse 40-42,
D–60486, Frankfurt-am-Main, Germany

Engelbert Zass
zass@chem.ethz.ch
Eidgenossische Technische Hochschule (ETH) Chemie-Bibliothek,
ETH Zürich, Universitätstrasse 16, CH–8092 ,
Zürich, Switzerland

Preface

In the fall of 1989 the ACS Division of Computers in Chemistry held a symposium on the newly released Beilstein Online database. From the success of and interest in this symposium, an ACS book, *The Beilstein Online Database: Implementation, Content, and Retrieval* (Symposium Series 436) was published in 1990. This book has been one of the best selling in this popular ACS book series. With the many positive changes that have occurred to the Beilstein Institute and database over these past eight years, interest in updating the Beilstein system grew to a point that has led to this book.

While this volume is, in one sense, an update of the 1990 ACS Symposium Series book (which is now out of print), in many ways it is a completely new book. Although the guts of the database have remained of the same high quality and value, the total computerization of all the information in the *Beilstein Handbook*, coupled with the development of a number of excellent software packages, has created a new Beilstein system. Additional new software packages (AutoNom and CrossFire) and new databases (NetFire, CrossFire Abstracts, and CrossFire Gmelin) have become important tools in the chemical community. These new products have been well received and used by chemists and other scientists in almost all disciplines (organic, analytical, physical, and inorganic chemistry; biochemistry; and others), both in universities throughout the world as well as by large and small industrial organizations, ranging from pharmaceutical companies to "ordinary" chemistry laboratories.

The goal of this book is to bring an understanding of the Beilstein databases and software to readers so they can easily learn both what information and data are available and how to obtain them.

This book is unique in the publishing history of the ACS Books Department. It is the first book to be completely prepared, submitted, and processed in electronic form by the ACS. All chapters were sent to me by FTP (file-transfer protocol), e-mail attachments, or disk. The final manuscript, over 50 million bytes including all figures, was sent to the ACS via FTP. Just as the transformation of Beilstein from a handbook-based product to an all-electronic one has created a powerful and cost-effective product, the all-electronic publishing of this book will, I hope, lead to new electronic books and journals that will become available to the chemical community faster, be better (in terms of multimedia and other content), and cost less than the old print equivalents.

In the preface of the previous ACS Symposium Series book on the Beilstein Online database, I wrote: "Beilstein Online is a renaissance or reawakening of a sleeping giant." This book again shows how Beilstein has reinvented itself and developed into an even more giant resource for organic chemistry, as a result of the many scientific, technical, and administrative changes to the Beilstein system.

I thank Clemens Jochum of Beilstein Information Systems in Germany for his encouragement on this project. Besides acknowledging the authors of the chapters in this book for their very considerable talents and efforts, I thank Susan Carino of Beilstein Information Systems in the United States for all her efforts in reading and commenting on the chapters.

Stephen R. Heller
Silver Spring, MD
May 1997

Chapter 1

The Beilstein System: An Introduction

Stephen R. Heller

This chapter presents recent developments of the Beilstein database and its component system parts, along with an overview of the many topics, databases, and software systems covered in this book. A short description of the Gmelin database is also presented.

This book is designed to update the reader on the status of the Beilstein database and search system and derived electronic information, as well as provide the chemistry community with a comprehensive and up-to-date view of the overall activities and scope of the current Beilstein system. When the first ACS book on Beilstein was published in 1990,[1] there was a renewed and growing interest in Beilstein as it moved from its past environment of a print-based German language reference, existing only as a very large series of books. This series made up the *Beilstein Handbuch der Organischen Chemie*, or, in English, the *Beilstein Handbook of Organic Chemistry*. In the short time since 1990, the staff of Beilstein (first in the Beilstein Institute and now in Beilstein Information Systems) have continued their vast modernization program, which has resulted in many new chemistry data and information resources for the chemical community. This book is a tribute to Konrad Beilstein (Figure 1.1), the man who had the foresight to realize that high-quality scientific chemical data and information would be a timeless resource for chemists.

Introduction to the Beilstein System

Throughout the history of abstracting scientific literature in chemistry, the name Beilstein has had a unique position of quality and value to the chemist. As a young graduate student in organic chemistry, I first came across Beilstein as part of my synthesis work on a bicyclic nitrogen ring system. In 1963 it took me a considerable amount of time to search for this class of compounds and translate the appropriate sections of the *Beilstein Handbook* into English; now those activities take a few seconds, because the database is computerized and the information is essentially

© 1998 American Chemical Society

Figure 1.1. Friedrich Konrad Beilstein.

all in English. The transformation of this sleeping giant into a modern information resource is amazing and of great value to the ongoing progress of most academic and industrial chemistry.

Table 1.1 gives a brief chronology of Beilstein—the man, the handbook, and the database. It is easy to see that the period between the first ACS book and this book was a time of major development and change for the Beilstein database and search system.

Even though the *Beilstein Handbook of Organic Chemistry* covers *only* organic chemistry, it has been a critical resource for most of the 20th century. As we approach the 21st century, the management and leadership of Beilstein Information Systems GmbH has further developed this valuable resource into a number of practical, everyday tools for the chemist. The purpose of this book, the successor to the ACS Symposium Series

Table 1.1. Beilstein Chronology

1838: Friedrich Konrad Beilstein was born in St. Petersburg, Russia.

1865: Beilstein became Professor of Organic Chemistry, St. Petersburg.

1866: Beilstein became the Chair of Chemistry at the Imperial Technical Institute in St. Petersburg.

1881: First Edition of the *Beilstein Handbook* (2 volumes, 1500 compounds, 2200 pages).

1885: Second Edition (3 volumes, 4080 pages).

1906: Third Edition (8 volumes, 11,000 pages).

1906: Death of Friedrich Konrad Beilstein.

1918: Fourth Edition (German Chemical Society, Berlin).

1984: Fifth Edition in English (480 volumes, 400,000 pages).

1988: Beilstein Online on STN.

1989: Beilstein Online on DIALOG.

1990: ACS Symposium Series book: *The Beilstein Online Database: Implementation, Content, and Retrieval.*

1995: CrossFire—Beilstein file in-house with ca. 6,000,000 organic compounds with properties data.

1996: CrossFire Gmelin—Gmelin file in-house available with ca. 1,000,000 inorganic and organometallic compounds with properties data.

1997: ACS book: *The Beilstein System: Strategies for Effective Searching.*

book entitled *The Beilstein Online Database: Implementation, Content, and Retrieval,* is to demonstrate how the Beilstein database and associated software products have been evolving over the last 7 years since the first book was published.

With all the changes, remodeling, and reorganization of Beilstein in the past few years into two separate organizations, the Beilstein Institute[7] and Beilstein Information Systems[7] (explained in further detail later in this chapter), it is probably best to first answer the question, "What is Beilstein now?" Perhaps the best way to begin to answer this question is to show what Beilstein was in the past. Beilstein is an enormous set of reference books (Figure 1.2) that, although very well organized, occupied a great deal of space. It was relatively difficult to quickly locate information spread throughout the five different series composing the handbook. To find out what Beilstein is today, one need only search a number of Internet resources. A recent search showed many citations of the word *Beilstein.* For example, the Lycos[2] search engine retrieved almost 600 records. Other search engines, such as AltaVista[3], produced some 200 citations, whereas InfoSeek[4] produced 76 citations and WebCrawler[5] produced 59 citations. Lastly, Yahoo![6] found Beilstein in just one citation, the Beilstein Web site. These citations included a number of references to the Beilstein World Wide Web (WWW) site on the Internet, www.beilstein.com, as well as many chemistry department WWW sites around the

Figure 1.2. The current *Beilstein Handbook,* from the Basic Series to Supplements I–V.

world that have either the *Beilstein Handbook* or access to Beilstein in electronic format. There are also references to Beilstein the town in Germany, Beilstein jade, the Beilstein mountain in Austria, and so on.

Although there are many answers to "What is Beilstein now?", this book will cover only those that relate to the areas of chemistry for which the Beilstein Institute and Beilstein Information Systems are involved. When the first ACS book was published in 1990, it was easy to say that Beilstein was the handbook and the online database. Today, a mere 7 years later, the name *Beilstein* represents the handbook; Beilstein Online; Cross-Fire; CrossFire*plus*Reactions; and CrossFire Gmelin, Autonom, Current Facts, and more.

Evolution of the Beilstein Institute

To create this new world of the 21st century Beilstein, a new administrative approach was developed and implemented by the two presidents of the Beilstein Institute during the early 1990s, Clemens Jochum and Reiner Luckenbach. The evolution of the Beilstein Institute from a world-class institute to a world-class modern organization involved a considerable change in approach, attitude, and internal and external operations. Beilstein has now evolved and remade itself into a commercial venture, and it is run as a business, in a most businesslike manner. The transformation of all staffing and financial activities, from outsourcing the creation of the

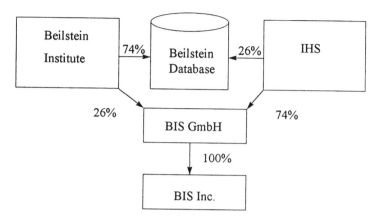

Figure 1.3. The new Beilstein corporate structure.

database, to the termination of German government support, to having all services paid for by users (a critical goal for the success of the Beilstein system) has produced a completely new Beilstein. Virtually nothing but the name and high quality are the same after this massive reorganization effort.

Beilstein has also changed administratively from being only the Beilstein Institute to being both the Beilstein Institute and Beilstein Information Systems. Beilstein, which was funded for years by the German publisher Springer-Verlag and the German government, is now changed, and a private company, Information Handling Systems (IHS), is responsible for the finances. A new corporate structure, shown in Figure 1.3,[8] shows these overall relationships. In this new structure, the database is now 74% owned by the Beilstein Institute and 26% owned by IHS. The marketing company, Beilstein Information Systems GmbH, is 74% owned by IHS and 26% owned by the Beilstein Institute. Lastly, the American subsidiary, Beilstein Information Systems, Inc., is 100% owned by the German company, Beilstein Information Systems GmbH. The new Beilstein database is now marketed and sold by IHS, a highly successful information company, which chose Beilstein as its first project in the scientific electronic database area. The Beilstein Institute staff now only continues to produce the *Beilstein Handbook*.

Lastly, software such as AutoNom (the automatic structure–nomenclature program) has been developed. It is from all of these changes and additions to Beilstein that the title of this chapter, *The Beilstein System,* was chosen. Because it is now a total system made of many parts and activities, it seems to be the most appropriate term to use.

Chemical Abstracts and Beilstein

One other point needs to be stressed before going further. The Beilstein system is primarily a system of data, which is extracted and processed information. Chemical Abstracts Service, a part of ACS, has produced *Chemical Abstracts* (CA) since 1907. CA is another valuable and critical source of chemical information.

Beilstein and CA are often thought of as competitors by some in the field of chemical information. In reality, the CA and Beilstein databases are complementary tools. CA covers virtually all of the chemical literature: organic, inorganic, physical, analytical, polymers, materials, and so on. CA also primarily covers the literature from a different and valuable perspective by providing an abstract and a summary that is a synopsis of each article. In comparison, Beilstein primarily covers the chemical literature focused on organic chemistry, dating back as far as 1771. Beilstein covers the chemical literature in which scientific data is presented, providing all the data presented in each published article as given by the original authors. Even though Beilstein has covered *classical* chemistry for most of the 19th and 20th centuries, as it moves into the 21st century it has responded to the needs of the chemical community and expanded its coverage of the literature to include toxicological and physiological effects of chemicals. Both CA and Beilstein have their audiences, which can overlap but are often quite different, and the applications of these two resources are also quite different. To obtain complete information, the research chemist must search both files to find all the hard data and other information on any compound of interest.

NetFire

Two potentially important topics are covered in a cursory way in this book, because they are too new for proper presentation and discussion. The first is NetFire, the new Internet-based current awareness organic chemistry literature search service from Beilstein. Released in December 1996, and made available to the chemical community for testing and evaluation at no cost for the first half of 1997, there is insufficient experience with NetFire to include it here. There has been one short description of NetFire written.[9] Basically, NetFire is a database of journal titles, authors, and abstracts from the organic chemical literature from 1980 to the present. The NetFire database contains the titles, abstracts, and authors of journal articles published in over 140 of the top journals in organic and medicinal chemistry. The Internet WWW query-by-form mechanism permits the formulation of a great variety of inquiries. You may search by author, by words included in the abstract, or by title (or words therein). You may also restrict your search to a certain journal or time range. Also included in the output is a cross-reference number to CrossFire, for those who have access to the entire Beilstein database.

CrossFire Gmelin System

The second topic not covered in any detail in this book is the CrossFire Gmelin system, which is the Gmelin database provided in-house under the Beilstein CrossFire software. The Gmelin database, produced by the Gmelin Institute of Frankfurt, Germany, is the most comprehensive collection of factual data on organometallic (coordination) compounds, alloys, glasses, ceramics, minerals, and physical chemistry. At present, the Gmelin database contains more than 1 million chemical substances and over 900,000 reactions reported in the literature from 1772 to the present and Gmelin handbook citations from 1772 to 1975. Gmelin contents include approximately

* 470,000 coordination compounds
* 55,000 alloys
* 14,000 glasses and ceramics
* 3200 minerals

With the Gmelin database as part of the in-house Beilstein CrossFire system, the two most comprehensive factual databases covering the areas of organic, inorganic, and organometallic compounds are now readily available under a speedy, convenient, and user-friendly client-server-based software system.

Coverage of the Periodic Table

Before outlining the details of this book, it is constructive to describe Beilstein's and Gmelin's coverage of the periodic table of the chemical elements. In Figures 1.4 and 1.5, the chemical elements included in Beilstein and Gmelin are presented.

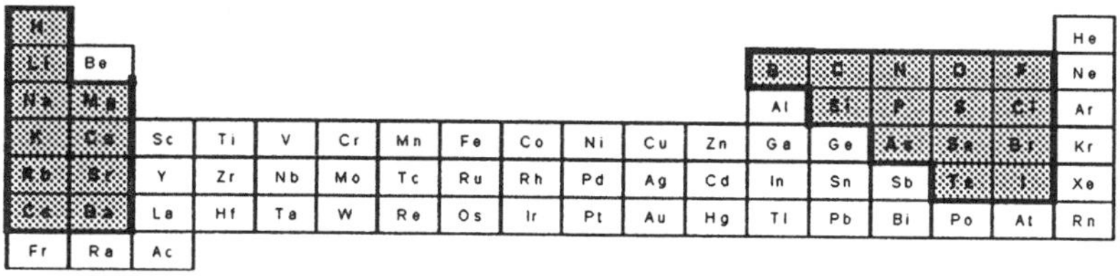

Figure 1.4. The periodic table showing which elements are covered in the Beilstein database.

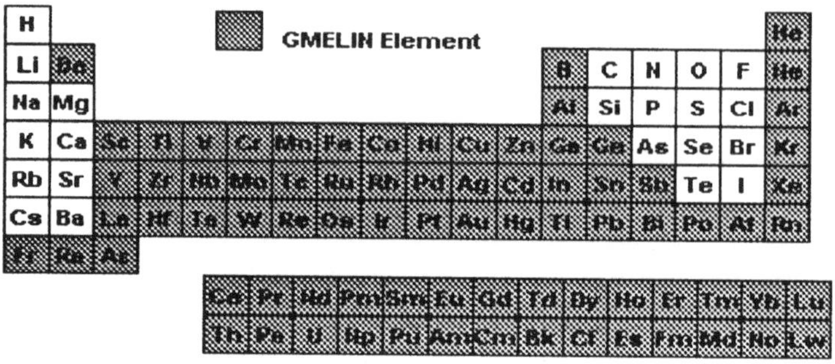

Figure 1.5. The periodic table showing which elements are covered in the Gmelin database.

Chapter Overviews

Chapter 2

Chapter 2, written by the recently retired Reiner Luckenbach, who was the president of the Beilstein Institute during the very exciting and turbulent times of the massive evolution of Beilstein into its current organizational structure, gives the background of the original *Beilstein Handbook*. The data quality control efforts are also described. This chapter helps put into perspective all of the other chapters in this book. It has been the massive massaging and manipulation of the existing *Beilstein Handbook* contents, coupled with the careful addition of additional extracted data and information, which has produced the evolving Beilstein system for the 21st century.

Chapter 3

Even though the Beilstein system has many new features, which are described in later chapters of this book, the Beilstein Online database, one of the original components of the Beilstein system, is still a very valuable resource for chemists. Andreas Barth, of STN-Karlsruhe, who worked with the Beilstein Institute in the late 1980s to develop the first version of the online database, has revised and updated his original contribution from the ACS Symposium Series book. This chapter includes information on the general structure of the information in the database and in each chemical compound record. A variety of search examples for data, physical properties, chemical reactions, and other information are given. Thus, Chapter 3 serves as the basic reference point for most of the information contained in this book.

Chapter 4

Wendy Warr, a well-known expert consultant in chemical information working with Bernd Wollny at Beilstein Information Systems, describes the Current Facts database, its content, its search capabilities, and its value to the chemist. Current Facts on CD-ROM is the Beilstein database of structures, data, and literature citations, designed to be used on a PC to provide recent information on chemicals reported in the literature. Current Facts contains 1 year's worth of information, updated quarterly, with the most recent 3 months of information replacing the oldest 3 months. The database goes back to 1990. One nice feature of this product is the structure searching capability built into it.

Chapter 5

Chapter 5 begins the treatment of the new Beilstein for the 21st century. This chapter, "Computer Systems for Substructure Searching", was written by one of the designers and developers of CrossFire, Dirk Walkowiak of Softron GmbH, in conjunction with John Barnard, a chemical information consultant and an expert in structure handling and search systems. The chapter starts with a history and introduction to chemical structure searching. From there the specifics of the CrossFire structure search system (general architecture, search algorithms, structure coding, and reaction retrieval) for the in-house Beilstein system are described in this chapter for the first time. This chapter describes the way in which CrossFire, which currently runs on the IBM RISC System/6000 computer system with a Unix operating system (other platforms, such as DEC and Windows NT, will be available in the near future), allows for very rapid searching of the more than 7 million chemical structures in the Beilstein database.

Chapter 6

After the treatment of structure searching and CrossFire in Chapter 5, Alexander (Sandy) Lawson of Beilstein Information Systems discusses CrossFire*plus*Reactions in Chapter 6. An improvement in the Beilstein system has been the extraction of reaction information, which is the heart of the *Beilstein Handbook of Organic Chemistry*. One of the critical and unique features of Beilstein has been the fact that virtually all data reported in the chemical and related literature for a chemical substance are entered into the system. All compounds entered into the Beilstein database have actually been synthesized and the structure has been assigned unambiguously.

This chemical approach of having a database comprising actually synthesized chemical substances (as opposed to the abstracting approach of CA) is one of the basic features of Beilstein that makes it so valuable to the bench chemist. In CA, published papers are abstracted on the basis of their being published in the chemical literature that CA covers (some 13,000 journals) plus patents. Even though the bulk of this material is the same information as is found in both the *Beilstein Handbook of Organic Chemistry* and database, there are differences. CA is a much larger database because its coverage goes beyond organic compounds and includes inorganic chemicals. Some chemicals that do not exist are in the CA database for two main reasons. Either they are chemicals that are being studied in theory or in computer analysis and calculation programs, or they are needed by the CA indexing system to allow for the proper indexing of derivatives (e.g., salts and parent ring systems) of a chemical. (This indexing should not be viewed as negative, because the CA indexing system and index database are one of its hidden virtues and a very valuable resource.) CA, owing to its broader coverage of the chemical literature, also has more reaction information in many cases for chemicals from more recent publications.

The combination of the reaction information content of Beilstein, which goes back to the 18th century, and the powerful new Beilstein Commander makes the new Beilstein product, CrossFire*plus*Reactions, a valuable and unique resource. CrossFire*plus*Reactions is a database and search system that should be a part of every synthetic organic chemistry research laboratory.

Chapters 7 and 8

Even though the chemical industry is the major user of the various Beilstein products, the chemist's first exposure to Beilstein comes when he or she is in school, either as an undergraduate or graduate student. Chapters 7 and 8, written by chemical information and library experts from leading academic institutions in the United States and Europe, Englebert (Bert) Zass (Chapter 7) and Ken Rouse and Roger Beckman (Chapter 8), describe how Beilstein is being taught and used in the university setting both in Europe and the United States.

In Chapter 7, Bert Zass (ETH, Switzerland), provides an excellent comparison and discussion of a number of reaction databases that are available both online and in-house, with emphasis on the Beilstein reaction database and the Beilstein CrossFire*plus*Reactions system.

In Chapter 8, Ken Rouse (University of Wisconsin) and Roger Beckman (Indiana University) describe a consortium of universities that have joined forces to make the Beilstein database available to a large community of students and academics via the in-house CrossFire system. In addition to discussing the very positive response from the academic-user community, the authors talk about the economics and finances of the system.

Chapter 9

From this background of the CrossFire system and examples of use in academia, Wendy Warr, in Chapter 9, discusses and explains its practical everyday value to the industrial chemist, and in particular the pharmaceutical chemist. Training, pricing, integration of information sources, NetFire, and CrossFire Gmelin are all covered in this chapter.

Chapter 10

Chapter 10 describes AutoNom, a software program that gives the chemical names of structures following International Union of Pure and Applied Chemistry (IUPAC) nomenclature rules. This chapter, written by the developer of AutoNom, Janusz Wisniewski of Beilstein Information Systems, is a detailed description of the performance of this program. The algorithmic approach that was used is also discussed in detail. AutoNom, which stands for AUTOmatic NOMenclature, is a program every chemist who has ever published a manuscript with a chemical structure will love. Chemists can easily draw structures, but few can name a structure according to the extensive and often complicated IUPAC or ACS/CAS naming rules. Instead of referring to structures as I, II, III, . . . , XL, AutoNom gives IUPAC names for organic and inorganic chemical structures. For companies that need to register substances under their proper chemical names in order to obtain approval from government and regulatory bodies, AutoNom is a very helpful tool. This chapter describes version 2.0 of AutoNom, the classes of organic and inorganic compounds it covers, and its limitations.

By the time you reach the end of Chapter 10, you will be both well versed in what Beilstein is today and is evolving into in the future, as well as knowledgeable about how the various Beilstein products can be of invaluable assistance to the everyday activities of almost every chemist.

References and Notes

1. *The Beilstein Online Database: Implementation, Content, and Retrieval;* Heller, S. R., Ed.; ACS Symposium Series 436; American Chemical Society: Washington, DC, 1990.
2. The Lycos Internet address is: http://www.lycos.com.
3. The AltaVista Internet address is: http://altavista.digital.com.
4. The InfoSeek Internet address is: http://guide.infoseek.com.
5. The WebCrawler Internet address is: http://webcrawler.com.
6. The Yahoo! Internet address is: http://www.yahoo.com.
7. Both the Beilstein Institute and Beilstein Information Systems are located in Varrentrappstrasse 40-42, Carl-Bosch Haus, Frankfurt (Main) 90, D–60486 Germany.

8. This figure is adapted from a talk by Bob Massie, CAS, at the Herman Skolnik ACS Award symposium for Reiner Luckenbach and Clemens Jochum, Chicago, IL, August 1995.

9. Heller, S. R. *Trends Anal. Chem. 16,* **1997,** p 112. The Internet address for this article is http://www.elsevier.nl:80/inca/homepage/saa/trac/frames.shtml

Chapter 2

The *Beilstein Handbook*

Reiner Luckenbach

The Beilstein Handbook of Organic Chemistry *is unique among handbooks of organic chemistry in that it provides a collection of critically examined and carefully reproduced data on the known organic compounds. In this respect, Beilstein is superior to all other normal bibliographical documentation and series of abstracts. Moreover, the* Beilstein Handbook *is the world's most extensive collection of physical data on organic compounds in printed form.*

In the constantly expanding world of chemical information systems, the word "Beilstein" has always been regarded as synonymous with high quality, reliability, and comprehensiveness. To maintain these criteria, a number of quality control mechanisms are applied at all production stages involved in the creation of the Beilstein data pool from which all Beilstein products are derived. These mechanisms include the application of man-ual (intellectual) data selection processes as well as a number of sophis-ticated automatic checking methods for each piece of data. Consequently, the quality and reliability of all Beilstein information tools are assured.

Introducing Beilstein

By the mid-19th century, organic chemistry had become more and more elaborate and mature, so that the amount of knowledge in this discipline began to increase dramatically. The *Beilstein Handbuch der Organischen Chemie* was the answer to the growing need for information for the organic chemists of that time. The first edition of this systematic collection of all carbon compounds known at that time (around 15,000) was published between 1881 and 1883 in two volumes on a total of 2200 pages.

Friedrich Konrad Beilstein, the editor of the handbook, was born in St. Petersburg in 1838 and studied in Heidelberg, Munich, and Göttingen. In 1866, he accepted an appointment at the Imperial Technological Insti-tute in St. Petersburg (following Mendeleev), where he taught, and, at the same time, edited and published the handbook. The third edition of the handbook was in print when F. K. Beilstein died in 1906. He had devoted 40 years of his life to editing it.

© 1998 American Chemical Society **13**

The fourth edition of the *Beilstein Handbuch der Organischen Chemie* (now generally called by its English name, the *Beilstein Handbook of Organic Chemistry*, and commonly referred to as "Beilstein") has been published continually since 1918. In its present state in 1996, it consists of more than 470 volumes comprising over 270,000 pages.

Introduction to the *Beilstein Handbook*

The *Beilstein Handbook of Organic Chemistry* contains important information about the preparation and properties of carbon compounds published since 1779. It contains reported research data (taken from over 2000 primary journals, patents, books, dissertations, and conference proceedings), which have been critically sifted, assessed in light of current chemical knowledge, and presented in a reliable and logical form. The editors carefully point out errors in published data, direct attention to the doubtfulness of some published statements, and check assertions of a speculative nature against the most recent findings. To organic chemists and all other scientists working in related fields (such as biochemistry, analytical chemistry, inorganic chemistry, physical chemistry, and others), Beilstein is an indispensable source of information. The alert user will be able to avoid embarking on false trails and be stimulated in further research by making use of this unique and valuable source of information and data.

The complete work is divided into series covering the periods listed in Table 2.1. Although the Basic Series and Supplementary Series E I through E IV are in German, Supplementary Series E V, publication of which started in 1984, is now exclusively in English. Series E V incorporates important new features (e.g., the use of Chemical Abstracts Service Source Index (CASSI) abbreviations for literature citations). The acceptance criteria for data have been sharpened in E V, resulting in an even more compact work with an extremely high density of accurate information.

Table 2.1. The Series of the *Beilstein Handbook*

Series	Abbreviation	Period of Literature Completely Covered	Color of Label on Spine
Basic Series	H	Up to 1910	Green
Supplementary Series I	E I	1910–1919	Dark red
Supplementary Series II	E II	1920–1929	White
Supplementary Series III	E III	1930–1949	Blue
Supplementary Series III/IV	E III/IV[a]	1930–1959	Blue–black
Supplementary Series IV	E IV	1950–1959	Black
Supplementary Series V	E V	1960–1979	Red

[a] Volumes 17–27 of Supplementary Series III and IV, covering the heterocyclic compounds, are combined in a joint issue; Series H to E IV are bound in brown, Series E V is bound in blue.

Each of these series comprises 27 volumes (or groups of volumes) in which the individual compounds are arranged according to the Beilstein System, which classifies all organic compounds according to their structures. A more detailed description of the Beilstein System and its application to finding compounds in Beilstein can be found elsewhere.[1-4] Which compounds are dealt with in which Beilstein volumes are shown in Tables 2.2 and 2.3.

For each chemical compound described in Beilstein, the following aspects are covered:

- structural formula
- compound name(s)
- molecular formula
- constitution and configuration (with particular regard to the clarification of these aspects, the most recent literature findings up to the date of publication of each subvolume are used)
- natural occurrence
- isolation from natural products
- preparation and purification (biochemical methods of formation are included when these are suitable for use on a preparative scale)
- structural and energy parameters of the molecule, including data on:
  conformation
  conformer equilibrium
  energy difference between conformers
  bond length and angles
  electron distribution
  dipole moment
  quadrupole moment
  polarizability
  coupling phenomena
  molecular deformation
  molecular potentials
  dissociation energy
  ionization potential
- physical properties, including data on:
  description of the physical state
  description of mechanical properties
  transport phenomena
  energy data
  optical properties
  spectral data
  magnetic properties
  electrochemical behavior
  physical properties of multicomponent systems

Table 2.2. Main Divisions of the Beilstein Handbook

Main Division	Volume Nos.	System Nos.
Acyclic compounds	1–4	1–449
Isocyclic compounds	5–16	450–2358
Heterocyclic compounds	17–27	2359–4720

Table 2.3. Contents of the 27 Volumes of the *Beilstein Handbook*

Beilstein Volume No. for Main Division — A (Acyclics), B (Isocyclics), C (Heterocyclics: Type and number of ring heteroatoms).

Type of registry compound	Feature of the functional group	A (Acyclics)	B (Isocyclics)	C: 1O*)	C: 2O*),3O*),...	C: 1N	C: 2N	C: 3N,4N,...	C: 1N,1O*),1N,2O*),...,2N,1O*),2N,2O*),..., further heteroatoms**)
(1) Compounds without functional groups	–		5			20			
(2) Hydroxy-compounds	–OH	1	6	17			23		
(3) Oxo-compounds	=O		7			21	24		
	=O + –OH		8						
(4) Carboxylic acids	$\gtrless$O/OH ; [$\gtrless$O/OH]$_n$	2	9						
	$\gtrless$O/OH + –OH; $\gtrless$O/OH + =O; $\gtrless$O/OH + =O + –OH	3	10						
(5) Sulfinic acids	–SO$_2$H								
(6) Sulfonic acids	–SO$_3$H								
(7) Seleninic acids, Selenonic acids, Tellurinic acids	–SeO$_2$H and –SeO$_3$H –TeO$_2$H	4	11	18	19	22	25	26	27
(8) Amines	–NH$_2$		12						
	[–NH$_2$]$_n$; –NH$_2$ + –OH		13						
	–NH$_2$ + =O; –NH$_2$ + $\gtrless$O/OH; –NH$_2$ + ...		14						
(9) Hydroxylamines and Dihydroxyamines	–NH–OH –N(OH)(OH)								
(10) Hydrazines	–NH–NH$_2$		15						
(11) Azo-compounds	–N=NH								
(12) Diazonium compounds	–N≡N]$^{\oplus}$								
(13) Compounds with groups of 3 or more N-atoms	–NH–NH–NH$_2$, –N(NH$_2$)$_2$, –N=N–NH$_2$, etc.								
(14) Compounds containing carbon directly bonded to P, As, Sb, and Bi	e.g. –PH$_2$, PH–OH, –P(OH)$_2$, –PH$_4$,..., –PO(OH)$_2$		16						
(15) Compounds containing carbon directly bonded to Si, Ge, and Sn	e.g. –SiH$_3$, –SiH$_2$(OH), ...								
(16) Compounds containing carbon directly bonded to elements of the 3rd–1st A-groups of the periodic table	e.g. –BH$_2$, –BH(OH), ..., –Mg$^{\oplus}$								
(17) Compounds containing carbon directly bonded to elements of the 1st–8th B-groups of the periodic table	e.g. –HgH, –Hg$^{\oplus}$,...								

*) instead of O also S, Se, Te (cf. p. 25)   **) e.g. B, Si, P, but not S, Se, Te (cf. p. 25)

- chemical properties (reactions) (in this section, reactions considered to be of sufficient importance are described for each compound with all types of reacting species)
- characterization and analysis
- salts and addition compounds

Altogether, about 420 data types (including more than 350 kinds of physical data) are found in the *Beilstein Handbook*.

Arrangement of the *Beilstein Handbook*

Apart from its division into series (*see* Table 2.1), Beilstein is arranged into compound classes (e.g., acyclic, isocyclic, and heterocyclic), as shown in Table 2.2. Even though the series (H, E I, E II, . . .) are devoted to the literature of consecutive time periods, the classification of the subject matter in each of the supplementary series is the same as that chosen for the 27 volumes of the Basic Series. Consequently, any particular volume of each supplementary series always contains the same classes of compounds (and only these) as the volume of the Basic Series with the same number. This correlation of subject matter, of benefit to the user, necessitated a further subdivision of volumes of E III, E III/IV, E IV, and E V, which are several times more extensive than those of the Basic Series. In addition to the 27 volumes of text, volumes 28 and 29 contain a general subject index (index of compound names) and a general formula index.

The additional volumes 30 and 31, published in 1938 and reserved for the naturally occurring polyisoprenes and the carbohydrates, have been discontinued. Their subject matter has since been incorporated into volumes 1–27 according to a constitutional formula following the general rules of the Beilstein System.

The classification of compounds into compound classes and the arrangement of the individual compounds within these classes are determined by the Beilstein System.[1-4] This system consists of rules by means of which *each and every* carbon compound can be assigned one specific location within the overall set of all carbon compounds. The total set is subdivided by 4720 system numbers that can be used for the purpose of orientation within Beilstein. A summary of the contents of the 27 volumes of the *Beilstein Handbook* is presented in Table 2.3.

Arrangement of Individual Beilstein Entries

First of all, the heading of each entry or article contains the most important names in boldface type; furthermore, names in spaced type are considered by the editors as less suitable (usually on account of poor compatibility with official nomenclature rules).

Following the empirical formula and the constitutional formula, the headings may further contain a back reference. A back reference such as "H 93; E II 64; E III 190" in a volume of Supplementary Series E IV means that previous records or entries (from a previous series) on this compound are to be found in the same volume of the Basic Series on page 93, of Supplementary Series E II on page 64, and of Supplementary Series E III on page 190. The prefix "vgl." (the English equivalent is "cf."), sometimes encountered in front of the back references, indicates that in the earlier records or entries the compounds prepared may have had a different configuration.

The lack of a back reference implies that the compound concerned has not been described in earlier series. In such a case, it is often possible to locate related compounds with the aid of either the coordinating reference (which indicates the page of the Basic Series on which the compound would have been described had it been known then) or the system number. These indicators, together with the series and volume numbers (system numbers only from E V, vol. no. 23/1) are to be found at the top of each odd-numbered page.

Pages of different supplementary series with identical coordination references or system numbers thus contain items on compounds with the same or closely related constitutions. At the top of each even-numbered page of the handbook is the designation of the compound class to which the compounds dealt with on that page belong. The text of the articles is set out in standardized form according to content, which is sorted into main subject groups. A list of main subjects is found in the section "Introduction to the *Beilstein Handbook*".

How To Search For and Find Compounds

Centennial Index

The Centennial Index, published in 1991 and 1992, probably represents the most efficient means of accessing information from the older (Basic Series through E IV) handbook series. Encompassing the entire literature period 1779–1959 (the 272 subvolumes with a total of more than 200,000 pages comprising the complete German language part of the *Beilstein Handbook*), the Centennial Index enables users to search the first 180 years of organic chemistry by using a single reference work.

Additionally, where important new information on a given compound has come to light since publication of its E IV (or E III/IV) entry, the Centennial Index indicates this by citation of its corresponding entry in E V (or even its post-E V original literature citations).

The Centennial Index is divided into two sections:

* General Subject Index (General Sachregister, volume 28, 10 volumes, approximately 15,100 pages)
* General Formula Index (General Formelregister, volume 29, 13 volumes, approximately 16,500 pages)

The Centennial Index supersedes the General Indexes H–E II, which covered the Basic Series and the first two Supplementary Series.

All names in the Centennial Index are formulated according to the International Union of Pure and Applied Chemistry (IUPAC) rules as used in E IV. Names found in the older series have been modernized to bring them in line with current IUPAC conventions. The names used in the index ("Index Names") differ from the names used in the text in that (1) substitution prefixes and saturation-state prefixes are placed after the index stem name, and (2) all stereochemical descriptors are omitted. This ordering means that all stereoisomers fall under one entry. The Centennial Index contains the names of more than 1.2 million compounds.

The names of addition compounds and salts whose cations are metal ions, complex metal ions, or protonated bases do not appear in the index. Information about such substances will be found at the end of the entry for the organic compound involved.

One interesting fact that one finds from this index is that with 545 pages (printed in double columns), comprising more than 41,000 compounds, "Essigsäure" (acetic acid) constitutes the most extensive single entry in the Compound Name part of the Centennial Index. This means that 2.7% of all compounds reported in the literature during the first 180 years of organic chemistry were "acetates"!

Other Indexes

The E IV cumulative subject indexes and cumulative formula indexes (green and black labels), forerunners of the Centennial Index, are similar in concept to the Centennial Index, with the exception that they were provided for individual volumes (e.g., volume 1) or groups of volumes (e.g., volumes 17 and 18), so that it was necessary to ascertain, by using the rules of the Beilstein System[1-4] or the computer program SANDRA (described in the next section), the volume in which the compound of interest would be located.

Indexes relating to the E V series are also arranged by compound class, corresponding to particular groups of volumes (oxygen heterocycles, for example, are listed together in the Collective Index for Volumes 17–19). These indexes, similarly, are divided into Compound Name Indexes and Formula Indexes.

The Formula Index is usually the surest and quickest method of searching for an individual compound. For searches of wider scope, such as for a compound plus its derivatives, the subject index (Compound Name Index) is better suited but requires either an adequate knowledge of nomenclature or the use of Beilstein's AutoNom software tool (*see* the subsequent discussion and Chapter 10). The subject indexes contain not only the names occurring in the item headings but also compound names appearing in spaced type in the text of the item.

Each individual Beilstein (sub)volume also contains its own individual subject index and (from E III onward) formula index for that (sub)volume. At the beginning of each first subvolume (e.g., 27/1), users find—among other things—a section on "stereochemical descriptors", which gives an extensive overview of the stereochemical conventions as used throughout the handbook (with examples).

SANDRA Program

The most straightforward method for quickly locating a specific compound in the *Beilstein Handbook* is probably provided by the computer program SANDRA (Structure and Reference Analyzer). SANDRA, which incorporates all the logical but complex rules of the Beilstein System, constitutes a powerful and useful software package, which enables the user to draw the structure of the compound of interest by using a fast graphic input system. It then analyzes the structure and identifies, in most cases within a few pages, which part of the handbook should deal with the compound of interest or its relatives.

The output provides the following information:

1. volumes or subvolumes of Series E III, E III/IV, E IV, and those volumes of E V in print, and
2. information relating to the page headings to enable the user to quickly determine where the registry compound should be found. These are
   - system number,
   - H-page numbers (often a range: these numbers are the coordinating references described previously),
   - principal functional groups,
   - degree of saturation, and
   - number of carbon atoms in the registry compound.

Mechanisms of Quality Assessment and Quality Control as Applied to the *Beilstein Handbook*

The Beilstein name is synonymous with an integrated and quality-oriented multimedia information system. For more than 110 years, Beilstein has provided chemists with the world's most comprehensive collection of structures, properties, data, and associated information on organic compounds.

Although the Beilstein database contains virtually all information found in the primary literature, the publication of the printed handbook, for obvious reasons, must be restricted to a concise extract of the most reliable ("best available") data. One—if not *the*—cornerstone of the Beilstein philosophy has always been the perpetuation and perfection of Beilstein's primary goal: "critical evaluation of the published material".[5–9] In this connection, critical evaluation (scientific screening or scientific scrutinizing) means the competent judgment of information with respect to its validity (i.e., its compatibility with the generally accepted state-of-the-art), its novelty, and its overall scientific importance. Some of the most crucial steps of this quality-improvement procedure are outlined in the following sections.

Scheme I summarizes the production flow at Beilstein. Let us examine the most important individual production steps mentioned in this scheme with respect to their potential for quality assessment and quality improvement.

Primary Literature Excerption

Primary literature excerption for Beilstein is carried out essentially by subcontractors commissioned by the Beilstein Institute, who guarantee the quality of all delivered electronic excerpts. The completeness of those excerpts and their quality are controlled by the Institute.

Error-checking by a set of relevant programs has been built into the excerption program, and they are applied by the excerptor as well as by Beilstein's internal scientific staff. Checking includes:

- ascertaining whether the CODEN year and volume number of the excerpted journal are in agreement,
- correctness of the Chemical Abstracts Service (CAS) Registry Numbers,
- matching of molecular formula and structural formulas, that is, of the molecular formula given in the original publication, and the one calculated by the program from the structural diagram. If a mismatch occurs, then the input cannot be continued before the necessary corrections are made.
- checking properties involving ranges ("from . . . to"): is the second value greater than the first?
- checking whether all relevant parameters are given (e.g., "temp" for "heat capacity"). If not, the input mask cannot be left.

To give the reader a flavor of the kinds of data and information that are examined and checked by the relevant quality control software programs,

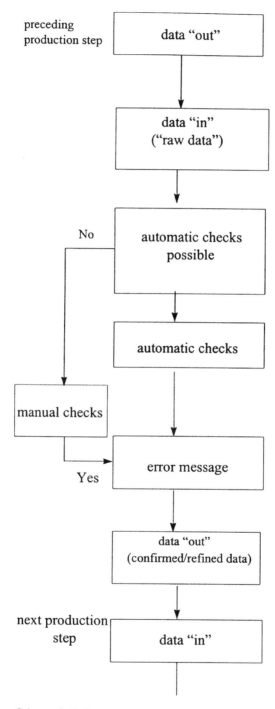

Scheme I. Beilstein production flow (overview).

Table 2.4. Some Examples of Error Messages During the Excerption

Error Message Number	Text (indication)
11	Non-numeric input in numeric field
13	Value out of range
15	Numeric field: more than 3 exponential digits
19	Numeric field: sign of exponent must precede the first digit of exponent
23	Numeric field: missing digits behind decimal point
96	Molecular formula: illegal electronic charge
102	Molecular formula: unknown characters or missing parentheses
131	CODEN: incomplete CODEN. Full length must be given
142	Molecular formula: illegal use of roman numerals; formulations such as "(I)" and "(III)" are not allowed to specify valence

Table 2.4 lists some of the possible 170 error messages that the input program may create and that the excerptor must consider for the correction work.

At Beilstein, various additional error-checking programs are applied, some of which check

- inconsistent stereochemical details for a compound,
- abnormal valences (e.g., pentavalent carbon),
- suspicious starting material–product relationships,
- inadmissible characters in the molecular formula,
- derivative for characterization without properties, and
- false nuclei in the NMR mask (the reported isotope is not present in the structure).

In addition, check programs control the realistic limits of certain properties (*see* Table 2.5 for some examples).

About 5–8% of all input from the excerptors is manually checked at random by the Beilstein Institute's internal control staff. Moreover, diskette material is directly compared with the relevant original literature. All subcontractors are contractually obliged to maintain a certain quality standard. In our experience this system has worked extremely well. Most data delivered by our subcontractors are of the highest quality.

After incorporation of the excerpted and—where necessary—corrected structures, data, and properties into our primary files, names are provided by AutoNom, the automatic naming program (*see* Chapter 10). A computerized Beilstein Ordering Algorithm developed from the SANDRA program is then applied to all structures, thus determining unequivocally the place of each individual compound in the *Beilstein Handbook*.

Table 2.5. Ranges of Physical Properties (Excerption)

Property	Range From	To	Unit
Optical rotation	−3,000	+3,000	degree
Bond moment	0	10	debye
Boiling point	−170	+500	°C
Melting point	−200	+500	°C
Critical density	0.1	5	g/cm^3
Dipole moment	0	50	debye
Enthalpy of vaporization	8,350	126,000	J/mol
Fluorescence maximum	150	900	nm
Refractive index	1	2	—
Dynamic viscosity	0	100	Pa·s
Vapor pressure	10^{-5}	10^4	Torr

However, the systematization of structures according to this system is more than just a necessary means for arranging compounds in the handbook. Because the system brings chemically related compounds into close proximity, the system is an indispensable prerequisite for the scientific data evaluation that significantly contributes to the quality assessment of the information as presented in the *Beilstein Handbook*.

Fine-Data Production

According to Scheme II, a proven and well-examined combination of automatic and intellectual (manual) checks is applied to all of the following fine-data production steps. This process leads to the clarification of dubious results, the verification of reported data, the precise assignment of stereochemistry, and so on. Additional programs at the control stages also check whether

- the compounds named have been duplicated,
- compounds ending in "-ium" have a positive charge,
- all sets of brackets in the compound name are paired,
- the numbering corresponds to the actual number of substituents in the names (e.g., 2,4,6-trinitrophenol).

Diagnostic messages are generated by the automatic error-checking programs. Action resulting from these messages is then taken in the following manual step within the fine-data production.

Although numerous automatic plausibility checks facilitate the quality assessment procedures at Beilstein, the competent scientist is nevertheless indispensable. His or her qualified evaluation finally leads to the

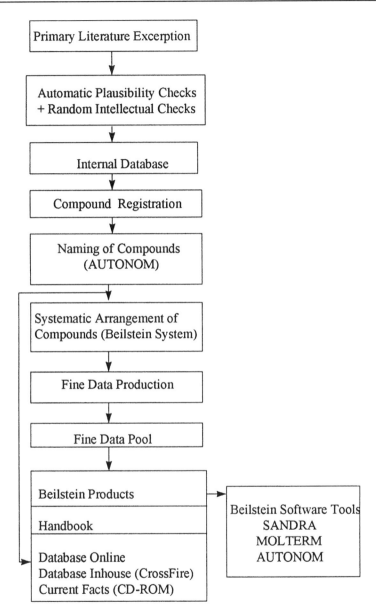

Scheme II. Beilstein fine-data production steps.

inclusion, correction, or sometimes even rejection of published data. This result is especially true for so-called "multipublication compounds" (i.e., entries in which the data and information come from a number of separate publications), where thousands of pieces of data must be evaluated.

On the basis of their scientific expertise, Beilstein's chemists are able, for example:

- to filter out multiple publications with the same principal content, well-known results, and trivial findings,
- to take into account corrections to earlier findings and to transfer them to similar cases,
- to recognize, and, where possible, clarify conflicting data on a given subject,
- to correct structural assignments of compounds, and
- to correlate individual results with analogous cases and reveal errors.

Some of the criteria checked during the intellectual screening and data evaluation procedure are

- material correctness of information (consistency with current general scientific principles),
- depth of information,
- completeness of information (are all relevant parameters given?),
- objectives of the publication,
- accessibility of (primary) information (type of journal, language, and date), and
- origin of publication.

The practical application of these and other relevant intellectual selection criteria to the processing of the *Beilstein Handbook* entries is exemplified in references 5–7 and 9. After completion of the critical evaluation and quality improvement steps described previously, the resulting data will either complement or replace the original data.

Stereochemistry

Assignments, corrections, and verifications of (doubtful) stereochemical features are areas that require the particular attention and expertise of the Beilstein Institute's editorial staff. Uncertainties relating to the configuration of chemical compounds, detected in the primary literature, are cleared up by further literature research or by analogy reasoning. Consequently, numerous stereochemical assignments found in Beilstein have not been given in the primary literature or elsewhere. This type of clarification is possible when one considers, for example:

- genetic relationships between a compound with configurational uncertainty and compounds whose configurations have been established,

- synthesis of the compound by methods analogous to those leading to compounds of known configuration,
- degradative reactions whose steric courses are known with certainty and that lead to the formation of the compound in question, and
- comparison of physical data within a homologous series.

The net result of this critical processing by the Beilstein Institute's editorial staff leads to a correction or verification of the constitution or configuration of almost every 10th compound reported in the primary literature, containing stereochemically relevant features.

Conclusions

The advantages and benefits for the user of this kind of data processing and the resulting quality improvement of chemical information result in

- improved confidence in the accuracy of the sifted data,
- greatly reduced risk of false conclusions and nonoptimal experimental planning as a result of inexact literature data,
- intensive intellectual reexamination of selected data becoming unnecessary (greater time saving), and
- promotion of further innovations due to the cross referencing of published information.

It should be evident that the data quality assessment of primary information by means of a science-based and intelligently executed evaluation program requires very high personnel and financial commitments. Despite these costs, Beilstein will strive to maintain its reputation of publishing only high-quality data.

Summary and Future

The current Fifth Supplementary Series was started with the publication of volume 17, that is, the first volume describing heterocyclic compounds. This decision was based on the results of extensive inquiries among users that indicated a clear preference in favor of these compounds. Heterocycles will go through volume 27. With more than 300,000 compounds, this will be, by far, the most comprehensive single Beilstein volume published in the history of the handbook. Its publication will continue through 1997 and will be highly facilitated by the introduction of additional computer assistance in the generation of the handbook text. The documentation of the acyclic and isocyclic (carbocyclic) compounds has been staggered and will be undertaken as soon as the heterocycles have been finished. The publication of E V will be complete by the year 2000.

It should be emphasized that the *Beilstein Handbook of Organic Chemistry*, apart from being an invaluable information tool in its own right, constitutes the indispensable and vital basis for all present and future electronic products. Without the handbook and its immense impact on generations of chemists in particular, and on the world of chemical information in general, the "electronic Beilstein", which is so knowledgeably described in the following chapters, would simply be nonexistent.

References

1. "How To Use the Beilstein Handbook of Organic Chemistry", 1995 (brochure, obtainable free of charge from Beilstein Information Systems, Varrentrapp-strasse 40-42, D-60486 Frankfurt am Main, Germany).
2. Luckenbach, R. *CHEMTECH* **1979**, *9*, 612–621.
3. Sunkel, J.; Hoffmann, E.; Luckenbach, R. *J. Chem. Educ.* **1981**, *12*, 982–986.
4. Luckenbach, R. *Acual. Chim.* **1985**, *8*, 39–42.
5. Luckenbach, R.; Ecker, R.; Sunkel, J. *Angew. Chem.* **1981**, *93*, 876–885; *Angew. Chem. Int. Ed. Engl.* **1981**, *20*, 841–849.
6. Luckenbach, R. *J. Chem. Inf. Comput. Sci.* **1988**, *28*, 94–99.
7. Luckenbach, R.; Sunkel, J. *J. Chem. Inf. Comput. Sci.* **1989**, *29*, 271–278.
8. Banciu, M. D. *Roum. Chem. Quart. Rev.* **1994**, *2*(1), 3–24.
9. Luckenbach, R. *J. Chem. Inf. Comput. Sci.* **1996**, *36*, 923–929.

Chapter 3

The Beilstein Online Database

Andreas Barth

The Beilstein database is the largest collection of critically reviewed factual information in organic chemistry. Since its first release the database has grown considerably, and currently it comprises over 7 million organic substances. The database contains chemical substance identification information, about 70 physical properties, chemical reaction information, and the corresponding literature references. STN International is an online service that offers very powerful features that are especially suited for a factual database such as Beilstein.

The Beilstein Online database is a structure-oriented factual database in organic chemistry containing critically reviewed and evaluated data from various sources.[1-3] Currently, the four sources of information for the database are

1. the printed handbook series from the Basic Series up to the fourth Supplementary Series (H through E IV),
2. the unreviewed excerpts from the primary literature until 1979 (short file data),
3. the fifth Supplementary Series, and
4. the material from the primary literature from 1980 onward.

All the information from the handbook has been thoroughly checked for errors and redundancies and has been published in the Main Series and the first four Supplementary Series. In addition to this information, there is a large collection of unreviewed excerpts on paper cards available covering the time span from 1960 to 1979, the so-called "short file". This material contains only the substance identification, an indicator of the type of factual information, and a reference to the literature. All information from the paper cards has been added to the database. In the meantime the publication of the fifth Supplementary Series has started and the corresponding information is also added to the database, thus replacing the former short file data. From 1980 onward the excerpts from primary literature are directly input to the database and are thus available in machine-readable form from the beginning of the creation of the database.

© 1998 American Chemical Society

The first part of the Beilstein database of organic substances was introduced in December 1988 on STN International. It was the first time that a factual database containing such a large number of physical and chemical entities had become publicly available as an online database. Initially, the database contained the structures and factual data of approximately 350,000 heterocyclic substances from the handbook, covering the time span from 1830 to 1959. Since 1988 the number of organic substances has increased substantially; the database presently contains approximately 7.3 million organic compounds (as of April 1997). In the future the database will be updated in short periods until it is up-to-date with the current literature. It is expected that regular updates will occur when the database has reached this state.

Each organic substance is identified by a chemical name given in International Union of Pure and Applied Chemistry (IUPAC)-oriented nomenclature and a structure diagram. In addition, a large set of physical properties and chemical information is described together with the corresponding literature references. The scope of information may cover substance identification information, synthesis and reaction data, structure and energy parameters, state of aggregation, mechanical properties, thermodynamic data, transport phenomena, optical and spectral data, magnetic and electrical data, electrochemical behavior, and multicomponent system data. In analogy to the handbook, the documents in the Beilstein database are substance oriented, that is, all factual information is associated with a well-defined chemical substance and a corresponding structure diagram.

For the individual chemical substances, the number of associated factual data may vary significantly. The minimum information corresponding to a substance comprises the identification data and one physical or chemical entity. With each factual data field, there is at least one literature reference and sometimes an additional note giving further information. Statistics from the current database of 6,777,157 compounds show the following: The major part of factual data is preparation data (PRE, 80.3%), melting point (MP, 52.3%), boiling point (BP, 9.6%), and reaction data (REA, 8.6%).

General Design of the Database

According to the data structure of the Beilstein database,[4] there are several hundred search fields and more than 140 different display formats. As discussed previously, the type and amount of information that is available for a particular substance may vary significantly. To obtain an overview of the available fields for a given substance, the display format FA (field availability) can be used. In Figure 3.1, the table of contents for the substance codeine is shown. The Messenger command language used here is described elsewhere.[5]

```
=> SEARCH codein/CN

L1      3 CODEIN/CN

=> DISPLAY 3 FA

L1      ANSWER 3 OF 3 COPYRIGHT 1996 Beilstein

Field Availability

Code    Field Name                          Occur.

MF      Molecular Formula                   1
SY      Synonym                             1
FW      Formula Weight                      1
SO      Beilstein Citation                  1
LN      Lawson Number                       3
RN      CAS Registry Number                 7
NTE     Notes                               1
RSI     Related Stereo Isomers              42
SF      Stereo Family                       1
PRE     Preparation                         2
MP      Melting Point                       6
INP     Isolation from Natural Product      1
PRE     Preparation                         2
CTMS    Mass Spectrum                       1
```

Figure 3.1. Table of contents for the substance codeine.

In the first column of this table the display formats are given, the full name is shown in the second column, and the number of occurrences is displayed in the last column. In this case, we find that there are two occurrences of preparation. The number of occurrences is a direct indication of the number of different preparation methods for the substance. To display the data for the substance, one may simply use the codes from the first column of Figure 3.2. A display of IDE (identification of substance) and PRE is shown in Figure 3.2.* Any combination of formats, including combined (predefined) and custom formats, is allowed.

For a deeper understanding of the Beilstein database, the structure of a Beilstein document must be outlined. There are essentially four different information levels (*see* Figure 3.3). All the substance identification information composes the first level. It is actually associated with a registered Beilstein compound (title compound). The availability information is on the second level. This information comprises the search fields field availability (FA), property hierarchy (PH), controlled terms (CT), and controlled terms of multicomponent systems (CTM). The content of these fields indicates whether there is information available for a specific property. The factual information (i.e., numeric values of properties and reac-

---

* In this case IDE includes the fields BRN, MF, SY, FW, SO, LN, NTE, RN, and STR (structure). In order to save space, the chemical structure has been deleted in Figure 3.2 by the author.

```
=> DISPLAY 3 IDE PRE

L1 ANSWER 3 OF 3 COPYRIGHT 1996 Beilstein

Beilstein Reg. No. (BRN):        5768734  Beilstein
Molecular Formula (MF):          C18 H21 N O3
Synonym (SY):                    Codein
Beilstein Reference (SO):        6-27
General Comments (NTE):          Stereo compound
CAS Reg. No. (RN):               76-57-3; 509-64-8; 7235-41-8; 16206-70-5; 27067-72-7;
                                 64520-25-8; 70982-46-6

...

Preparation:
PRE
     Start:      BRN=5631965 3-O-Acetylmorphin, BRN=1420865
                 dimethylacetamidedimethyl acetal
     Time:       3 hour(s)
     Yield:      12.00 %
     Solv:       xylene
     Temp:       160.0 Cel
     ByProd:     BRN=5777248 8.alpha.-(Dimethylcarbamoylmethyl)-8,14-dihydro-6-
                 desmethoxythebain \34 percent of Input , BRN=5776724 6-O-
                 Acetylcodein \30 percent of Input
     Reference(s):
     1.   Fleischhacker, Wilhelm; Richter, Bernd, Chem.Ber., 113 <1980> 12, 3866-
          3880, LA: GE, CODEN: CHBEAM

...
```

Figure 3.2. Display of IDE and PRE (extract).

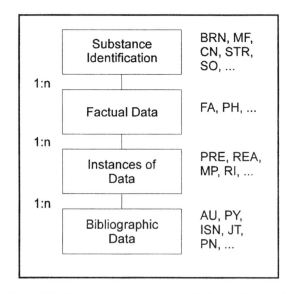

Figure 3.3. Document structure and information levels.

Field Name	Field Qualifier	Unit
Dipole Moment	DM	D
Temperature	DM.T	CEL(°C)
Method	DM.MET	-
Solvent	DM.SOL	-

Figure 3.4. Design of physical entities: dipole moment.

tion information) is found on the third level (measurement). On the fourth level the bibliographic information is given, referring to the measurements of the next higher level.

There is a one-to-many relationship between the higher and the next lower level in this hierarchy. This relationship means that for a given substance, there could be many properties available, for each property there could be several measurements, and for a measurement there could be more than one citation. A search could be performed in fields of different levels. In any case, the answer set will consist of Beilstein Registry Numbers (BRNs) and some additional information. This means that the result, if any, of a search is always a Beilstein title compound. The information about the search level can be reconstructed from the additional information in the answer set and is used for the various display formats. A certain value of the (P)-proximity is assigned to each measurement (instance of data), and the individual references are associated with a particular (S)-proximity value.

All information of the first two levels is indexed in standard search fields, such as CN (chemical name) or MF (molecular formula). The factual data have a slightly different structure because the main entity is, in general, dependent on further parameters. As an example, the data structure of dipole moment is shown in Figure 3.4. Here, the main entity is dependent on the parameters temperature, method, and solvent. The values for the dipole moment and the corresponding temperature are given in debyes and degrees celsius, respectively. The (P)-operator allows one to perform a search of the dipole moment at specific parameter values. The main qualifier (DM) is also used to display the data for dipole moment.

In addition to the custom formats consisting only of a single entity, there is a hierarchy of combined (predefined) display formats. These formats enable the user to display almost any amount of data using a single display format. Also, a dynamic display feature shows the customer the data related to his or her search query plus the substance identification information (QRD = query-related data). To display the content of a Beilstein document, the field availability (FA) is used (*see* Figure 3.1).

In the Beilstein database, the information is given in different forms, as textual data (free text), keywords, numeric values (ranges), and structures. Free text and keywords can be searched in the same way as in bibliographic databases. Numeric values are searched as ranges using the numeric relation operators, such as < (less than) or > (greater than). Chemical structures can be searched as exact structures or as substructures. They can be built either off-line using a graphical structure editor (e.g., STN Express) or online by using the Messenger STRUCTURE command.

Chemical Substance Information

The Beilstein database is a structure-oriented factual database, and thus the identification of substance (IDE) is the central information unit. All substances in the Beilstein database are identified by a sequential Beilstein Registry Number (BRN). It is the primary access key for the records (documents) in the file, and it can be used to directly show the information for a given substance without a previous search process. The individual components of a multicomponent system have their own Component Beilstein Registry Number (CBRN). Both numbers are searchable as numeric values, that is, they may be used in a numeric range search.

In addition, there is a chemical name (CN) and synonyms (SY), a molecular formula (MF) and related formulas, a formula weight (FW), and a structure (STR). All these fields are both searchable and displayable. Furthermore, there are many additional search fields generated from these input fields. Chemical names are given in IUPAC-oriented nomenclature. They are indexed as complete names in CN and as parsed segments in the fields CNS (chemical name segments) and BI (Basic Index). The segments are generated by using two different algorithms:

- parsing the names at all special characters such as a hyphen and comma, and
- applying a dictionary of natural segments developed by the Beilstein Institute.

Chemical name segments can be searched by using the well-known proximity operators (S), (W), and (A). An example of such a search is given in Figure 3.5.* In this case a search for derivatives of salicyl aldehyde is

---

* It must be noted that a chemical name segment search is not always complete because the parsing of chemical names does not generate a systematic and consistent set of name fragments.

performed. The answer set comprises 236 hits, and the 190th answer is also shown in Figure 3.5. A hit resulting from a search in the Basic Index may also stem from a starting material of a reaction or from a by-product, as was the case in this example. These names are also indexed in the Basic Index, but they are not necessarily registered as Beilstein compounds.

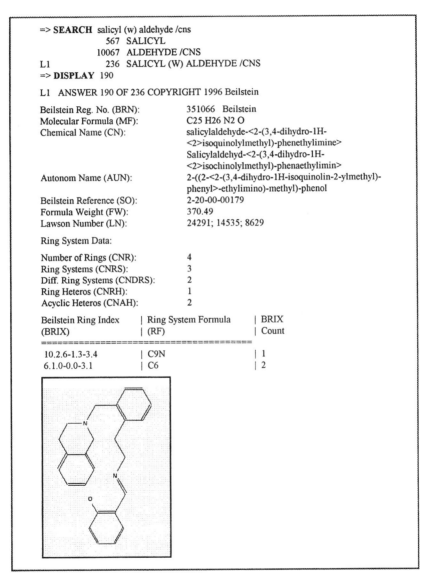

```
=> SEARCH  salicyl (w) aldehyde /cns
              567  SALICYL
            10067  ALDEHYDE /CNS
L1          236  SALICYL (W) ALDEHYDE /CNS
=> DISPLAY  190

L1  ANSWER 190 OF 236 COPYRIGHT 1996 Beilstein
```

Beilstein Reg. No. (BRN):	351066 Beilstein
Molecular Formula (MF):	C25 H26 N2 O
Chemical Name (CN):	salicylaldehyde-<2-(3,4-dihydro-1H- <2>isoquinolylmethyl)-phenethylimine> Salicylaldehyd-<2-(3,4-dihydro-1H- <2>isochinolylmethyl)-phenaethylimin>
Autonom Name (AUN):	2-((2-<2-(3,4-dihydro-1H-isoquinolin-2-ylmethyl)- phenyl>-ethylimino)-methyl)-phenol
Beilstein Reference (SO):	2-20-00-00179
Formula Weight (FW):	370.49
Lawson Number (LN):	24291; 14535; 8629

Ring System Data:

Number of Rings (CNR):	4
Ring Systems (CNRS):	3
Diff. Ring Systems (CNDRS):	2
Ring Heteros (CNRH):	1
Acyclic Heteros (CNAH):	2

Beilstein Ring Index (BRIX)	Ring System Formula (RF)	BRIX Count
10.2.6-1.3-3.4	C9N	1
6.1.0-0.0-3.1	C6	2

Figure 3.5. Search for chemical name segments and display of a substance record.

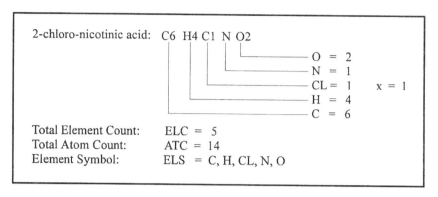

Figure 3.6. Generation of index terms from the molecular formula.

Searching on the molecular formula (MF) and the associated search fields is another possible way to identify a chemical substance. A number of additional search terms are generated from the molecular formula (*see* Figure 3.6). At first, the molecular formula is indexed in MF and BI. In addition, the single atom counts are generated for all chemical elements and some pseudoatoms such as X (halogen atoms) and M (metal atoms). For each element, the corresponding periodic group and element group are created. Furthermore, a total element count (ELC), a total atom count (ATC), and the element symbols (ELS) are indexed. The ATC and ELS fields are especially useful to limit the search to certain ranges of atoms or elements. The molecular formulas for the individual components are stored in the field component molecular formula (CMF). To search for the number of components of a multicomponent system, the field number of components (NC) can be used.

Chemical structures are the most important key for identifying substances in the Beilstein database. The user interaction and the substructure search capabilities are identical to those of the CAS Registry database and are described in the user documentation for the Beilstein database.[1] In Figure 3.7, a substructure search for derivatives of adenine is shown. By using PC-based software such as STN Express, it is also possible to build structures off-line and upload the connection tables into the database.

An important display format for the essential substance identification data of a set of answers is IDETAB, a tabular representation containing the answer number (ANS), the Beilstein registry number (BRN), the Beilstein preferred registry number (BPR), the molecular formula (MF), the field availability count (FA), and the total number of occurrences (OCC).

```
=> search  adenine/cn
L1              5  ADENINE/CN

=> display  5 brn

L1  ANSWER 5 OF 5 COPYRIGHT 1996 Beilstein
Beilstein Reg. No. (BRN):  5777  Beilstein

=> str  5777
:end
L2  STRUCTURE CREATED

=> search  l2 sss sam

...

L3         50 SEA SSS SAM L2

=> display  l3 4 qrd

L3  ANSWER 4 OF 50 COPYRIGHT 1996 Beilstein
```

Beilstein Reg. No. (BRN):	6335348 Beilstein
Molecular Formula (MF):	C11 H18 N6 O
Synonym (SY):	6-<N-<2-(N-ethoxy-N-ethylamino)ethyl> amino>purine
Autonom Name (AUN):	N,O-diethyl-N-<2-(9H-purin-6-ylamino)-ethyl>-
	hydroxylamine
Beilstein Reference (SO):	6-26
Formula Weight (FW):	250.30
Lawson Number (LN):	30692; 3634; 298

...

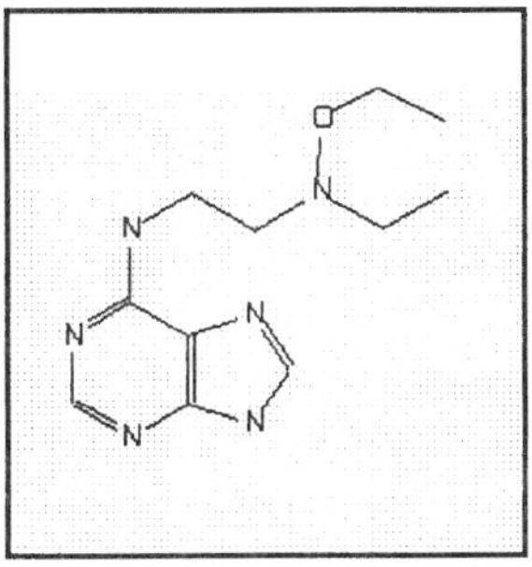

Figure 3.7. Substructure search for derivatives of adenine.

Chemical Reaction Information

Although the Beilstein database is not a typical reaction database, there is a large amount of chemical reaction information available. It is possible to find data on substance preparation, chemical behavior, and isolation from natural products (biosynthesis). Currently, all searches must be performed as text searches for chemical name segments or as searches for the availability of data. This is certainly a limitation for reaction searches.

```
=> search flavon?/cns and inp/fa
         1378  FLAVON?/CNS
        19693  INP/FA
L3          45  flavon?/CNS AND INP/FA
```

Figure 3.8. Search for the biosynthesis of flavone compounds.

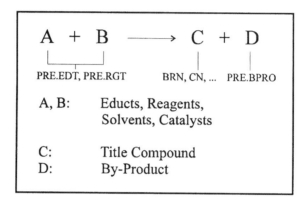

Figure 3.9. Reaction scheme for substance preparation.

In many cases, one retrieves the complete reaction information, including the literature references. Some cases, however, consist only of a reference.

By using the field availability (FA), one could search for the availability of reaction data. In the following example (Figure 3.8), we are interested in the biosynthesis of flavone derivatives. Information on biosynthesis could be searched in the field isolation of natural products (INP). Because we are not looking for any specific synthesis, we can search for "INP" in FA and combine this with the name segment "flavon?" in CNS (*see* Figure 3.8). The question mark ("?") stands for a right truncation of the search term.

The names of the substances taking part in the preparation of a compound are indexed in individual subfields of preparation (PRE). The reaction scheme for substance preparation is given in Figure 3.9. Here, a compound C is built from the starting materials A and B. The title compound C could be identified by BRN, CN, STR, etc., the starting materials are indexed in PRE.EDT or PRE.RGT, and by-products are found in PRE.BPRO. In the Beilstein database, one may also search for the chemical behavior (REA) of a compound. The corresponding reaction scheme is similar to the reaction scheme for preparation.

Notably, to an online searcher, where the reaction information is indexed in the Beilstein database is not always obvious. Sometimes, it is necessary to execute a query both in the preparation and the reaction fields. It is also necessary to keep in mind that only a subset of the Beilstein compounds have reaction data available.

Physical Property Data

The Beilstein database is the largest source of physical property data with respect to both the number of substances and the number of different physical properties. There are about 70 properties indexed in numeric fields and about 240 different keywords corresponding to physical entities stored in textual fields as controlled vocabulary. For all physical properties and controlled terms, there is at least one literature reference pointing to the source of information. In Beilstein the physical properties are conceptually ordered in a hierarchical manner, that is, like a thesaurus file.

Physical properties are measured or calculated quantities recorded as numeric values associated with an uncertainty and a corresponding physical unit. The numeric values including the uncertainty are stored as numeric ranges in the database. There is an implicit physical unit corresponding to each search and display field that can be changed by the customer. In general, physical properties also depend on a set of parameters such as temperature or pressure. With the STN Messenger software, it is possible to perform parameter-dependant searches of properties using a proximity operator.

The design of physical properties mimics the data structure (for an example, *see* Figure 3.4). There is an entity name, such as dipole moment, and a corresponding field qualifier, DM. The field qualifiers for the parameters are built by using the abbreviation for the entity, a dot (".") and an abbreviation for the parameter. Thus, the field qualifier for the temperature (T) of the dipole moment is DM.T. The main qualifier, such as DM, is also the display format for this entity. With each numeric field a unique physical unit is associated, that is, all values in a field are given in the same unit.

In the Beilstein database, most of the physical properties are given as numeric data. For historical reasons, some properties are recorded as exact values without an uncertainty. These properties are called *single-point entities*, and they can be treated as simple numeric fields in bibliographic databases, that is, only single values are stored. Physical properties that are recorded with an associated uncertainty are referred to as *numeric range entities*. A numeric range consists of a lower and an upper value that are equivalent to a value plus or minus an uncertainty. For these entities, the end points of the ranges are indexed in a numeric field.

The concept of *range searching* is associated with several different meanings, and it can refer to both the query and the indexed range. A sophisticated search feature has been developed to support the retrieval of physical properties in numeric databases. Properties that are recorded as a value plus or minus an uncertainty are difficult to obtain in standard public information retrieval systems that mainly focus on the retrieval of bibliographic (text) data. The standard numeric range searching capability of STN Messenger has been enhanced to enable the user to retrieve the *fuzzy* numeric data of physical properties. When the user performs a search of a fuzzy physical property, the Messenger software automatically invokes a procedure that performs an intersection between the query range and the stored (indexed) range. If the intersection is not zero, the corresponding document is retrieved as a hit. This feature is called a *numeric range overlap detection,* and it is actually based on an intersection between the query range (or point) and the stored ranges. The query could also be given as a single value, such as BP = 100. In this case, the software works in the same way as it does for a query range. The notation numeric range overlap detection refers to the overlapping of the query range (or single value) and the stored range. All numeric physical properties can be searched by using the standard numeric compare operators (=, <= or =<, <, >= or =>, >). In addition, the logical and proximity operators can be applied to numeric queries.

In Figure 3.10 a schematic representation of a search for an exact value is shown. Here, a boiling point (BP) of 100 °C is searched. This is represented as a straight vertical line in the figure. The stored ranges of our example are represented as horizontal intervals. Those horizontal intervals intersected by the vertical line compose the answer set, that is, the ranges 90–110 °C, 95–115 °C, 98–100 °C, and 100 °C. There are two exact indexed values, 100 °C and 105 °C. Of course, only the first value belongs to the answer set.

In Figure 3.11 a similar picture is given for a search of a numeric range. The range (96 °C ≤ BP ≤ 108 °C) is represented by the shaded rectangle. As in Figure 3.10, we obtain the answer set as a result of the intersection between the query range and the stored ranges. The answer set for this example includes the ranges 90–110 °C, 95–115 °C, 98–100 °C, 100 °C, 105 °C, 108–112 °C, and ≥108 °C. In this case, the stored open range (≥108 °C) is also part of the answer set.

In the previous searches, the answer set can become rather diffuse because of the broadness of the stored ranges, and it may be necessary to reduce the answer set by combining it with other factual information. However, the answer set is complete in the sense that no possible hit will be missed. Even an imprecise specification such as MP ≥ 50 °C will be found by every query overlapping with the indexed range from 50 to infinity; for example, a search for MP = 1000 °C will also retrieve this range.

Most physical quantities are associated with a unit serving as the standard measure for this entity. Even though much standardization has been done, several different unit systems are still in use. Of course, this

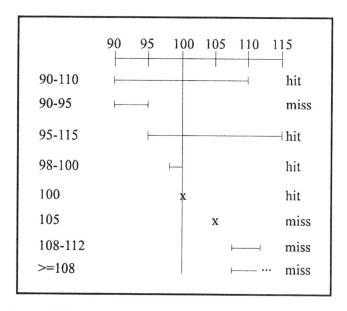

Figure 3.10. Schematic representation of a single value search.

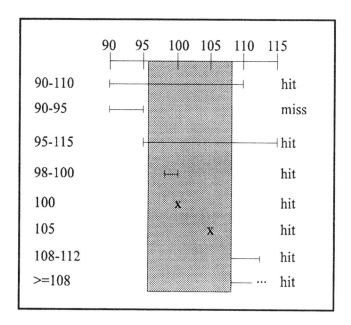

Figure 3.11. Schematic presentation of a numeric range search.

also depends on the area of application. In the Beilstein database most properties are measured in SI (Système International) units, but some quantities are still given in other units. To overcome the challenge of remembering all the units used in different numeric databases, STN has developed a feature for the conversion of units. This enables the customer to work in his or her preferred set of units independent of the units used in the file. The STN Messenger software will automatically do the unit conversion and search or display the data in the appropriate units. The unit conversion capability enables the user to do the following:

*   specify units with numeric search terms,
*   display the default unit for a property,
*   set the system unit for a numeric property according to his or her convenience, and
*   select a common standard for the system units.

In particular, the customer may work in the default units, globally set units, or overwrite a unit for a specific field. To illustrate this feature, a few examples are presented in Figure 3.12. In the first example, a search using the default units for the database is presented. If the customer formulates a query without specifying any units, then the default units are taken and no unit conversion is performed. Consequently, the display of this property shows the property in the original unit. In the second example, the explicit specification of the unit in the search statement overrides the default unit of this field. The numeric range is converted from the default unit into the unit of this search field. A subsequent display of this property shows the original units again. In the third example, the unit for heat capacity is globally set by the Messenger SET command. In this case, the unit is changed for the rest of the session unless the customer changes it again or overrides it explicitly. All subsequent searches and displays are now presented in the new unit. As for the SEARCH command, the user could also override the unit for a display of a property. This is done in the DISPLAY command shown in the last example in Figure 3.12.

As discussed at the beginning of this chapter, not all properties are available for each substance. In other words, there are holes in the file. This means that for a given substance, some properties are not present, either because they have not yet been measured or were not known at the literature closing date for the recording of the respective substance. When a customer is searching for a property with specific values, potential hits may be missed because some property values are not in the database, although they could possibly overlap with the search range. In a normal numeric search, these potential hits are not included in the answer set. However, in STN databases it is possible to search for these missing values or holes. In Figure 3.13 there is an example for a search of the holes in

```
Example: Default Units
=> search  120 - 140/cp
L1              421  120 J/MOL*K - 140 J/MOL*K /CP

Example: User Defined Units
=> search  0.05 - 0.08 kcal/mol*k/cp
L2              502  0.05 - 0.08 KCAL/MOL*K/CP

Example: Global Setting of Units
=> set  unit cp=kj/mol*k
SET COMMAND COMPLETED
=> dis  unit cp
CP          DEFAULT:          J/MOL*K
CP          CURRENT:          KJ/MOL*K
=> search  0.05 -0.1 /cp
L3              447  0.05 KJ/MOL*K - 0.1 KJ/MOL*K /CP
=> display  cp

L3   ANSWER 1 OF 447 COPYRIGHT 1996 Beilstein

Heat Capacity (CP):
Value              | Temp.          | Ref.
(CP)               | (.T)           |
(KJ/MOL*K)         | (Cel)          |
================================
0.09399 - 0.36794  | -213.1 - 26.9  | 1

Reference(s):
1.  Shakirov, R. F.; Andreeva, V. M.; Maslennikov, E. I.; Lyubarskii, M. V.,
    Russ.J.Phys.Chem.(Engl.Transl.), 55 <1981> 8, 1214-1215, LA: EN, CODEN:
    RJPCAR
    Zh.Fiz.Khim., 55 <1981>, 2138-2140, LA: RU, CODEN: ZFKHA9
```

Figure 3.12. Examples of using the unit conversion capability.

the field MP. This is done by combining the numeric search with the term "MP/FNA" (Field Not Available) using the Boolean OR operator. However, a simple search for missing values should not be performed because, in some cases, the number of holes is greater than the number of substances with values for this property and the system limits are easily exceeded. Thus, it is strongly recommended that the "FNA-Search" be used only if it is really necessary and that the search always be combined with some restrictions to keep the answer sets within the system limits.

In addition to the numeric search capabilities for the properties, it is possible to search for the availability of the property itself. For all properties that are available for a given substance, the property name is indexed in the field FA (field availability) and in PH (property hierarchy). Those properties having only a literature reference but no numeric value

```
=> search c=6 and (10 - 30/mp or mp/fna)
   209185  C=6
    19874  10 CEL - 30 CEL /MP
   6777157 ALL/FA
   3542220 MP/FA
   3234941 MP/FNA
           (ALL/FA NOT MP/FA)
L4          97704  C=6 AND (10 CEL - 30 CEL /MP OR MP/FNA)
```

Figure 3.13. Example for the retrieval of documents with missing values.

```
=> search  mpol/fa
L5              995  MPOL/FA
```

Figure 3.14. Search for the availability of physical properties.

in the database have an index entry both in CT/CTM (controlled terms/multicomponent system) and in PH. Hence, the three fields can be used for different purposes of availability searches:

• to retrieve any numeric value or a literature reference (PH),
• to retrieve any numeric value (FA), and
• to retrieve literature references only (CT/CTM).

Expanding FA provides the list of property names and the corresponding number of occurrences, that is, the number of substances for which the property is available. The strategy for searching for property names is very simple, and it corresponds to searching controlled vocabulary in bibliographic databases. An example is given in Figure 3.14. Here, we have searched for the availability of numeric data for molar polarization.

General Fields and Bibliographic Information

In addition to the entities described in the previous section, there are several fields with general content. First, there is the BI containing data as single words from all text fields. In BI one can search for chemical name segments or for general information such as toxicity data or ecological information. This field can be used in a similar way as in other files.

The factual data are always given together with one or more literature references. A part of this bibliographic data is also searchable in the file. Because Beilstein quotes only the surname of the first author, there is a limited way to search names in AU. The publication year is searchable in PY, the patent number is found in PN, the journal title is given in JT, and the corresponding CODEN is indexed in ISN (international standard number). The bibliographic information refers to an instance of a measurement of a property or a single chemical reaction (*see* Figure 3.3). This information corresponds to the lowest information level and not to a title compound as a whole such as the reference number in the source (SO) field. The (P)-proximity is used to refer to an instance of factual data. However, if two or more bibliographic search terms are combined in one query, the (S)-proximity must be used. The reason for this is simply that there may be more than one reference for a single instance of factual data.

Another type of information is represented by the descriptors in the fields FA, PH, preferred property names (PPN), and controlled terms (CT and CTM). All these fields contain information on the availability of properties in the form of bound phrases. If data are available for a certain field of a given substance, the field name and the corresponding field qualifier (search field code) are indexed in the field FA. In the case of properties the same information is indexed in PH. In addition, there may be descriptors present for those properties that are not stored as factual data in Beilstein. These descriptors are indexed in the controlled terms field (CT) or in the corresponding field for multicomponent properties (CTM). A customer may conduct searches in these fields to obtain information on the availability of certain properties or to reduce the size of an answer set. It is also possible to use FA as a starting point for a numeric property search.

Summary and Conclusions

The Beilstein database is the largest numerical database in terms of substances information, physical properties, and chemical reaction information. It is available on two online hosts: STN International and Knight-Ridder Information. For a comparison of the different implementations of the Beilstein database, the reader is referred to other works.[6–7] The access capabilities of STN Messenger provide the customer with powerful search and retrieval tools. In addition, it is possible to perform multifile and cross-file searches together with the most important databases in science and technology, such as the databases of Chemical Abstracts Service (e.g., CA, Registry, or CJACS), Derwent (e.g., WPI), Chemical Concepts (SpecInfo), ISIS (e.g., Science Citation Index), or many other database producers.

The rapid growth of the Beilstein database will soon result in a database that is up-to-date with the current literature. It will be possible to search factual information (physical properties and chemical reaction information) for the complete time span of chemical information and documentation in a single database. Hence, the Beilstein database will become one of the most important information sources in chemistry, physics, and related sciences.

Acknowledgment

The funding of this work by the Federal German Ministry for Education and Technology (formerly Research and Technology) is gratefully acknowledged.

References

1. Luckenbach, R. "The Beilstein Handbook of Organic Chemistry: The First Hundred Years," *J. Chem. Inf. Comput. Sci.* **1981**, *21*, pp 82–83.
2. *100 Jahre Beilstein 1881–1981* (booklet to the 100th anniversary of the Beilstein Handbuch), Beilstein Institute: Frankfurt (Main), 1981.
3. "Building a Structure-oriented Numerical Factual Database," in *Chemical Structures—The International Language of Chemistry*; Jochum, C.; Warr, W. A. (Eds.); Springer-Verlag: Berlin, 1988; pp 187–194.
4. *Beilstein Database Description Manual*, STN: Karlsruhe, Germany, 1993.
5. *A Guide to Commands and Databases*, STN: Columbus, OH, October 1988.
6. Buntrock, R. E.; Palma, M. A. "Searching the Beilstein Database Online: A Comparison of Systems," *Database* **1990**, *13*, pp 19–34.
7. "A Comparison of Searching the Beilstein Database on Different On-line Vendors," in *Proc. Montreux 1990 Int. Chem. Inf. Conf., Montreux, September 24–26, 1990*; Buntrock, R. E.; Palma, M. A.; Collier, H. (Eds.); Springer-Verlag: Heidelberg, 1990; pp 125–158.

Chapter 4

Current Facts in Chemistry on CD-ROM

Wendy Warr and Bernd Wollny

Current Facts in Chemistry on CD-ROM contains one year's worth of factual and numeric data for about 300,000 compounds in the organic chemistry literature. It is even more up-to-date than the CrossFire database. Structures, facts, and bibliographies can all be searched. The system runs on PCs under Windows.

A recent publication classified chemistry CD-ROMs under the headings patents, encyclopedias and full-text journals, bibliographic databases and related products, chemical catalogs and chemical names, health and safety, spectra and physical properties, and drug information.[1] Many titles were listed, but only a few offered chemical structure display, and fewer than ten featured structure and substructure searching.[1,2] The number of titles has perhaps doubled since 1994, but there are still only three chemical structure searching systems designed specifically for CD-ROM. The Softron Substructure Search System, S[4], is one of these.[3–5] This is the system used in Beilstein's Current Facts in Chemistry on CD-ROM, which was one of the first "real chemistry" products to become commercially available.[6–8]

The system works under Windows on a 386SX PC-compatible computer (or higher) with 4 MB of RAM, 5 MB of hard disk space, a Microsoft or Microsoft-compatible mouse, CD-ROM drive, MS-DOS version 3.1 or higher, Windows 3.1 or higher, and MSCDEX version 2.1 or higher. Any graphics printer supported by Microsoft Windows may be used. A networked version is available. There have been many complaints in the past about publishers who perpetuate the undesirable practice of expecting old CD-ROMs to be returned. Beilstein is not guilty of this: users who cancel their subscriptions are not asked to return disks already in their possession. The product contains two separate databases (facts and structures), and two sets of search software: the S[4] search software (Softron GmbH, Munich) and FULGOR software (Running Bytes GmbH, Berlin).

The Databases

The CD-ROM contains structures, data, and literature references for about 300,000 compounds per year. Each disk contains one year's (i.e., four quarters') worth of information. Publication is quarterly. Each update adds the latest quarter's information to the database and removes that of the oldest quarter. Current Facts is one quarter more up-to-date than CrossFire; the fourth disk of 1996 covered the literature from July 1995 to July 1996. At first, 84 journals were abstracted for Current Facts; after 1990 the number of journals was increased to about 120. Current Facts in Chemistry is just one part of the Beilstein Information System, and it can be used in conjunction with the Beilstein database proper. Thus, the Comments field in Current Facts tells the user whether further data on the compound is available in the online database.

The data structure has over 650 separate fields, of which 350 are searchable. Of these 350 fields, more than 70 are indexed numerically and more than 20 are indexed as string fields. There are 23 keyword fields in the database containing nearly 230 subjects. The data structure is inevitably complex, but for the specialized user this database is obviously of great importance; written and/or online help with the complexities is at hand.

The Structure Editor and Structure Searches

Current Facts has a user-friendly, state-of-the-art structure editor (*see* Figure 4.1). It takes little time to become proficient with it, although version 1 has a few quirks. The merge facility, for example, can give a valence error when naphthalene is created from two benzene rings, but it is still possible to do a successful search for naphthalene by ignoring the "error".

The Softron Substructure Search System, S^4, is a state-of-the-art system for fast searching of very large databases.[3-5,8] It is reputedly particularly suited to the technical problems of searching on CD-ROM.[3,8] Substructure searching uses the "free-site" concept. (Users more familiar with STN and ISIS tend to assume that substitution is possible unless otherwise specified, whereas the free-site method, as in Questel DARC and S^4, uses the opposite convention.) Generic group specification, tautomer search, and stereochemical discrimination in search are all featured. The most significant point to be mentioned, however, is speed. Searches are so fast that it is hard to believe that more than 300,000 structures are being processed. Although S^4 searches are exceptionally fast for tightly defined queries, anecdotal evidence suggests that it is possible to tie up the search for many minutes with certain "fuzzy" queries. However, this author has not personally experienced any slow searches. The substructure searched for is highlighted in each hit displayed.

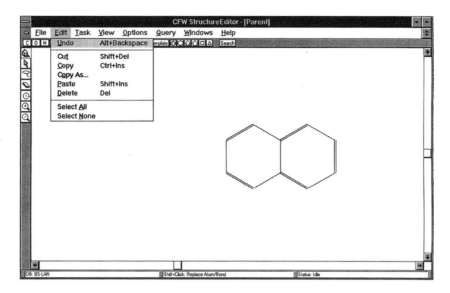

Figure 4.1. Structure Editor.

The Fact Editor and Text Searching

The search software is FULGOR (Running Bytes GmbH, Berlin). The edit screen has three sets of boxes, allowing Boolean operators to be easily selected, field names to be picked, and field values to be keyed in. There is also an extra box for the display and editing of the final search string. A proximity operator is featured. The data structure has over 650 separate fields, of which 350 are searchable, but the clever way of handling field hierarchies (*see* Figure 4.2) makes data searching relatively easy.

Queries can be stored and recalled, a search history for the current session can be displayed, and queries can be combined easily using the Fact Editor and selecting query numbers (say, q1 and q2) and logical operators. Indeed, previously retrieved hit lists, whether from structure or factual searches, can be intersected with either new factual searches or with other lists to refine the search further. An important feature of the system is the lack of limits on the size of hit sets, so that very large lists can be intersected if necessary.

The Windows clipboard can be used to advantage in the usual way. Also, fact queries can be copied to the clipboard in STN or DIALOG format for subsequent online searches. Special conversion routines have been written to allow copying to and from ISIS/Draw, STN Messenger, and DARC.

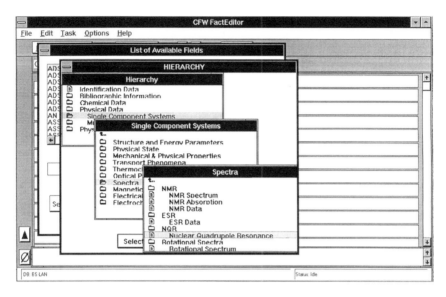

Figure 4.2. Field Hierarchy Windows in Fact Editor.

Displaying Results

There are three main display modes: short factual, short structure (Figure 4.3), and full display (Figure 4.4). Short mode (for structures or text, or structures and text) is used for displaying more than one hit per screen, with user-definable numbers of columns and rows. Any substructure searched for is highlighted in each hit displayed. It is possible to create subsets, selecting hits from the short display.

Data from a display can be copied into the Fact Editor via the clipboard. This feature can be used to advantage if, for example, the user sees an interesting piece of information on displaying a hit set and wants to retrieve all the compounds and information present for the reference in question. When data is copied from the display onto the clipboard, no matter how much of a fact is selected, the complete fact including all subfields is always copied. If necessary, subsequent search time can be reduced by deleting some of the subfields.

The hyperlinks feature is a powerful tool. Every Beilstein Registry Number (BRN) displayed in the text of a compound's record has been implemented as a hyperlink. If a user clicks on the highlighted BRN in a preparation field, he or she can instantly display the full record for the starting material. Moreover, by clicking on the hyperlinks in this new record, the user can find its starting material. Users can thus browse through whole synthesis paths.

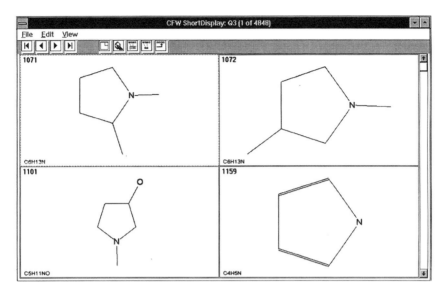

Figure 4.3. Short structure display.

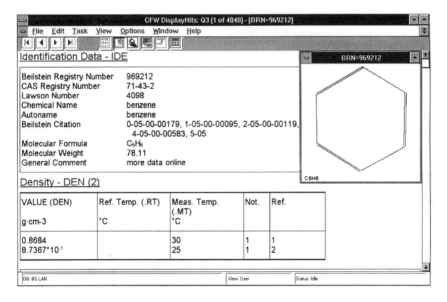

Figure 4.4. Full display.

This is a factual, numeric database, not just a bibliographic one (although it does have full and searchable bibliographies), so in some cases really detailed information can be printed out and displayed. For example, the compound numbered BRN 1073 produces a 2-1/2 page printout of structure, data, and literature references, including six tabulated moments of inertia, half a page of data on solution behavior, and so on. Printing is normally fast and efficient, but there are some problems with very large structures.

Documentation

A user guide describes the system in general and teaches structure building and searching. It is concise and accurate, and has a simple index. The user guide is a reference type of manual, with each feature described in order of occurrence in menus, rather than a teaching guide, but it can still be used as a learning tool. The system is quite intuitive, and reference to the manual soon becomes unnecessary.

The manual could be clearer in explaining Field Availability Search and in differentiating between "facts" such as melting point and boiling point, and "identification facts" such as molecular formula and chemical name. American users may be puzzled by being told to use "I", not "J", for iodine, but this is essential information to German users. All users should beware of different versions of chemical names (e.g., "thiophen" may occur as well as "thiophene") and of both British and American spelling.

Another manual is supplied entitled *Hierarchical Field Description*. This manual does not make for light reading, but it is an essential reference work for the user who needs to access the wealth of factual and numeric data, to check on the units used for the numeric data, and so on. For example, it tells the user that kinematic viscosity is a coefficient defined as the ratio of the dynamic viscosity of a fluid to its density, and that it is temperature dependent, so that an associated parameter field for temperature is needed.

Future Developments

The software features of Current Facts in Chemistry on CD-ROM have not been enhanced recently. This is a reflection of the uncertain nature of the long-term future for CD–ROMs. Not surprisingly, Beilstein Information Systems has decided to concentrate its development efforts on Cross-Fire. A new version of Current Facts in Chemistry on CD-ROM, planned for release at the end of 1997, will be integrated with the Beilstein Commander interface.

References

1. Warr, W. A. "Chemical Structure Handling on CD-ROM: Recent Developments and Future Prospects," in *Proceedings of the 18th International Online Information Meeting*; Learned Information: Oxford, U.K., 1994; pp 131–137.
2. Warr, W. A. "Chemically Intelligent CD-ROMs: Beyond Names and Images," in *Proceedings of the 1994 International Chemical Information Conference*; Collier, H., Ed.; Infonortics: Tetbury, Gloucestershire, U.K., 1994; pp 149–166.
3. Hicks, M. G.; Jochum, C. "Substructure Search Systems. 1. Performance Comparison of the MACCS, DARC, HTSS, CAS Registry MVSSS, and S4 Substructure Search Systems," *J. Chem. Inf. Comput. Sci.* **1990**, *30*, 191–199.
4. Barnard, J. M. "Structure Representation and Searching", in *Chemical Structure Systems*; Ash, J. E.; Warr, W. A.; Willett, P., Eds.; Ellis Horwood: Chichester, England, 1991; pp 9–56.
5. Warr, W. A. "Systems for Chemical Structure Handling," in *Chemical Structure Systems*; Ash, J. E.; Warr, W. A.; Willett, P., Eds.; Ellis Horwood: Chichester, England, 1991; pp 88–125.
6. Berks, A. H. "Beilstein Current Facts in Chemistry on CD-ROM," *J. Am. Chem. Soc.* **1992**, *114*(9), pp 3576–3577.
7. Heller, S. R. "The Beilstein Current Facts in Chemistry CD-ROM," *J. Chem. Inf. Comput. Sci.* **1991**, *31*, pp 430–432.
8. Hicks, M. G. "Beilstein Current Facts in Chemistry: A Large Chemical Database on CD-ROM," *Anal. Chim. Acta* **1992**, *265*(2), pp 291–300.

Chapter 5

Computer Systems for Substructure Searching

John M. Barnard and Dirk Walkowiak

Substructure searching is the heart of any organic chemical database of structures and related data and information. This chapter introduces the topic of substructure searching and then focuses on the specific way in which the Beilstein database is structured and searched using the CrossFire system.

Substructure searching is one of the most important computational problems in the design of chemical information systems, and a wide variety of different approaches and algorithms have been developed and used over the past forty years. Stated most simply, substructure searching is the process of identifying those members of a set (or database) of full structures that contain a specified query substructure. Because the conventional two-dimensional chemical structure diagram can be regarded as a topological graph, substructure searching is an example (and indeed, one of the most important practical applications) of the so-called *subgraph isomorphism* problem in graph theory.[1-3]

Effectively, testing for subgraph isomorphism involves trying to find a correspondence or "mapping" between the atoms (nodes) in the query structure (graph G_Q) and a subset of the atoms in a database structure (graph G_F) in such a way that the bonds (edges) of G_Q simultaneously map to a subset of the bonds in G_F. In other words, if two atoms in the query are joined by a bond, then they can correspond to two nodes in the database structure if and only if the two atoms in G_F are also joined by a bond; this is known as the *adjacency condition*. Furthermore, the relevant atom and bond types (node and edge labels) must be identical if the atoms or bonds are to correspond to each other. In practical terms, things are slightly more complicated than this, as the query may permit alternative or wild-card values for some of the atoms and bonds, but the principles remain the same.

The subgraph isomorphism problem is fairly well understood from a theoretical point of view, and it is known to be a member of a class of mathematically equivalent problems, called the NP-complete problems, for which there are no known algorithms whose worst-case time requirements do not increase exponentially with the size of the input (the number of

© 1998 American Chemical Society

atoms in the two structures being compared in this case). It is strongly believed by theoreticians (although it has not been rigorously proven) that there can be no algorithms for NP-complete problems that will always work in polynomial time[4] (i.e., in proportion to the square, cube, etc., of the number of atoms, rather than in proportion to a constant raised to the power of the number of atoms, or worse). This makes substructure searching an intrinsically slow process, and there are fundamental theoretical limits on the extent to which it can be speeded up.

However, things are not quite as bad as the theory may make them seem because the point about NP-complete problems is that, although the *worst-case* time requirements must always be exponential in the size of the input, algorithms can be found in which the *average* time requirements are acceptable. This is particularly true of the very simple graphs that are used to represent chemical structures, where, for example, different atoms have different numbers of connections (and few have more than two or three), and there is a reasonable variety of different atom and bond types that cannot correspond to each other. Even with a "brute force" algorithm (in which every conceivable mapping of query atoms to database structure atoms is tried in turn, either until one is found that obeys the adjacency condition—and a "hit" is identified—or until all of them have been tried without success), it is possible that by chance the very first mapping tried will be a hit and the algorithm can stop straightaway.

Finding subgraph isomorphism algorithms that operate with acceptable average time requirements has occupied the attention of developers for nearly 40 years, and it is still the subject of active research. Many of the published algorithms have found their way into operational systems, and the various systems with which the Beilstein database can be searched employ a number of different approaches. There are three main types of techniques that have been used, often in combination:

1. use of a faster computer, or several computers at once,
2. use of various heuristics to improve the chances of finding an isomorphism early on, or to reject candidates without exhaustive testing, where they cannot give rise to isomorphisms, and
3. carrying out the most time-consuming operations in a preprocessing of the database, which is independent of the query substructure.

The first of these is not in fact as flippant as it appears, and given the increase in computer power per unit cost that has taken place in the last four decades, it might even be considered to have been the most popular approach. Though most of the techniques used in the second approach were developed a long time ago, they still find wide application and continue to be refined and improved. The third approach was more recently developed and lies behind a number of systems from the late 1980s and early 1990s, including both S[4] and CrossFire.

Parallel Computing

A conceptually simple approach to speeding up the searching of a large database is to divide it into several portions and search each simultaneously on a different machine. The best-known example of this is the use of pairs of minicomputers in parallel for the substructure searching of the Chemical Abstracts Registry File on the STN International System.[5] An advantage of this approach is that as the database grows, further search machines can be added to maintain overall performance. Some research work has been done using more sophisticated approaches to "database parallel" searching,[6–7] though it has generally been found that the costs involved in controlling and supplying data to the various parallel processors soon outweigh the benefits, and such methods have not been used in operational systems. Similar problems were encountered in attempts to investigate "algorithmically parallel" approaches, in which the operations involved in some of the algorithms described in the next section were distributed over several processors.[8–9]

Backtracking and Partitioning

Backtracking is an example of the second type of technique described above, and it is the "traditional" method for the final "atom-by-atom" stage of a substructure search. The first such algorithm was published by Ray and Kirsch in 1957,[10] though algorithms based on the same principle continue to appear.[11–12] Backtracking operates by comparing each database structure in turn with the query, and either accepting it as a hit or rejecting it. In its simplest form the algorithm starts with an arbitrary atom Q_1 in the query Q, and maps it to an arbitrary atom F_1 (of the same atom type) in the file structure F. It then proceeds to map each neighbor (Q_2, Q_3, etc.) of Q_1 onto an unmapped neighbor (F_2, F_3, etc.) of F_1; if this is successful, the algorithm continues by trying to map each unmapped neighbor of Q_2 onto an unmapped neighbor of F_2, etc., until all the nodes in Q have been mapped, in which case a match has been found. If, at any stage, it proves impossible to find a mapping in F for an atom in Q, the algorithm "backtracks" to the last successfully mapped atom of Q, and tries an alternative mapping for it. If there are no further alternative unmapped nodes in F onto which the current node in Q might map, the algorithm backtracks again. If the backtracking gets back to the starting point, Q_1, and all the alternative possible mappings for Q_1 have been tried, the structure is rejected because there is no match.

Backtracking algorithms offer an improvement over brute force by not completing the mapping of all nodes in Q, once it can be seen that those nodes already mapped cannot form part of a subgraph isomorphism. The worst-case time requirements still increase exponentially with the number of nodes, but the design of the algorithm makes it unlikely that the worst case will arise very often. The performance of backtracking algorithms can be improved by using more detail than simple atom type (for example, the number of neighbors) when choosing possible correspondents, and by careful selection of the order in which alternatives are examined. For example, if an unusual hetero atom, with a lot of neighbors, is chosen as Q_1, it is unlikely that very many nodes will be mapped successfully (and thus potentially need to be unmapped as the algorithm backtracks) unless there is really a match.

A popular way of improving the performance of a backtracking algorithm is to use a partitioning procedure. This procedure is based on the division of the nodes of each graph into subsets of potential correspondents, which are then refined. The purpose of the partitioning is to reduce the number of possible mappings that must be investigated, and it is initially done by using some property of the nodes, such as atom type or number of connections. For example, if the query contains a node that is a nitrogen atom with connections to two other atoms, this can only correspond to file structure nodes that are also nitrogen atoms with connections to at least two other atoms. The initial partitioning is then refined by a process of further subdivision. In some cases this may leave certain query atoms without any potential correspondents in the file structure (or vice versa) in which case there cannot be a match at all, and the algorithm can terminate without having to do any backtracking.

The refinement step is, in many cases, based on a technique known as *relaxation*, in which the description of a node is enhanced by iteratively examining its immediate neighbors; thus, at each iteration, information from more and more distant nodes can be brought into the description of a particular node. Relaxation is a commonly used technique in chemical structure handling, though not often identified by name; the best-known example is the Morgan algorithm[13] for canonical numbering of the atoms in a molecule.

Several algorithms using this type of approach have been published over a long period. Most of them use a basic backtracking procedure as a backup to the partitioning procedure, and those that do not run the occasional risk of incorrectly identifying a substructure match when none in fact exists. Among the most important partitioning or relaxation algorithms are those due to Sussenguth,[14] Figueras,[15] Ullmann,[16] and von Scholley.[17] Though Ullmann's algorithm was published in 1976, it was only in the late 1980s that it began to be used in chemical substructure search applications, and studies by Willett and his students[18] have suggested that, for such applications, it may be the most efficient of all subgraph isomorphism algorithms published to date. One of its advantages is that it is readily able to identify *all* the isomorphisms present simultaneously (where the substructure appears multiple times in the same file structure), rather than terminating as soon as it finds the first one.

Backtracking-based atom-by-atom routines are used as the final search stage in many operational systems, including STN Messenger[19] (the "iterations" phase), Questel-Orbit's DARC System,[20] the conventional MACCS and ISIS systems from MDL Information Systems, Inc.[21] (though not MACCS FastSearch), and Hampden Data Services' PSIBase software,[22] which is used in microcomputer/CD-ROM implementations of databases such as Chapman & Hall's Dictionary of Drugs[23]. The two-dimensional substructure search component of Chemical Design Ltd.'s Chem-X software[26] specifically uses the Ullmann algorithm.[16]

Screening

Although the algorithms described in the last section offer considerable improvements over brute force, they remain time consuming and thus impractical for searching more than a few thousand compounds on conventional computers. Screening systems allow the bulk of the compounds in a database, which cannot match the query, to be eliminated rapidly from consideration, so that only a restricted number of candidates need to be examined with the rigor of a backtracking search. Screening systems normally involve indexing the compounds in a database by use of a set of search keys, each of which describes some structural feature (such as a small substructural fragment) of the molecule. A corresponding set of keys can be identified for the query structure, and those file structures that do not contain all the keys from the query can be eliminated from further consideration.

The choice of keys is, of course, crucial. It is a general principle of indexing systems that middle-frequency keys are the most useful: commonly occurring ones are of no use because they do not allow any structures to be eliminated, and very uncommon ones are of little use because it is unlikely that they will ever occur in queries. As far as possible, the keys must also be independent of each other, since if one key always or nearly always occurs with another, it contributes little to retrieval performance. The set of keys for a particular structure is often represented as a string of bits, where the presence of a structural feature causes one or more bits to be switched on; searching is then a simple logical comparison of bit strings (though it can be speeded up by "inverting" the bit map). In some systems, such as the substructure search system on STN,[19] a fixed dictionary of fragments is used, and structural features that are not included in the dictionary do not get represented in the bit string; the dictionary was established by a study of the statistical occurrence of different fragment types in a large file.[24]

In the Merlin substructure search system, developed by Daylight Chemical Information Systems, Inc.,[25] the structural features are short sequences of atoms and bonds, and the description of each fragment is directly used (via a *hashing algorithm*) to set particular bits in the bit string (or *fingerprint*). The shorter sequences set more bits, which allows some account to be taken of their relatively higher frequency of occurrence. In this system, a process of "folding" or "superimposition" can be used to compress a long bit string to form a shorter one (with consequent savings of storage space). So much space can be saved (with only 32 bytes or fewer being needed for each fingerprint) that on modern machines with large amounts of random-access memory, all the fingerprints for an entire database can simultaneously be held in memory; this allows for extremely rapid screen searches, leaving only a relatively small number of compounds that must be tested for matching using a backtracking procedure. A similar approach is used in the Chem-X system of Chemical Design, Ltd.[26]

Tree-Structured Fragment Searches

The most recent developments in new algorithms for substructure searching have been based on carrying out a query-independent preprocessing of the database to set up hierarchically structured trees essentially representing hierarchically organized substructural fragments. This work has some relationship to two systems, developed in the 1970s, which employ a hierarchical tree structure for substructural-fragment-based screening. In the Chemical Information System (CIS) developed at the National Institutes of Health,[27] there are separate searches for atom-centered fragments and ring-system descriptors. The fragment search proceeds from the central atom type, via the number of neighbors it has, to the atom type and bond orders for each neighbor in turn (see Figure 5.1); in the ring-systems search, successive levels of the search tree describe the ring pattern, the atom types present, the hetero atom positions, and the ring substitution positions. An atom-by-atom backtracking search is available for rigorous searching of those compounds retrieved by the fragment- and ring-probe searches.

In the DARC system,[28] the fragments (called FRELs, for Fragment Reduced to an Environment which is Limited) describe two concentric "layers" of atoms around a *focus*, which is an atom with at least three (or in some cases two) nonhydrogen neighbors. The FRELs generated from the structures in a database are also stored in a hierarchical tree, with generalized forms called *fuzzy FRELs* being included at higher levels of the tree. In searching, the FREL search (which uses only as many FRELs from the query as is necessary to reduce the number of candidate database structures to an acceptable number) is followed by a bit-string match, mainly concerned with ring systems, and a backtracking atom-by-atom match.

The Hierarchical Tree Substructure Search (HTSS) system was developed in Hungary[29] in the mid-1980s, and was at one time used for substructure search of the Beilstein database on the Maxwell Online (Orbit) system; this service was withdrawn, though, in September 1992. Each

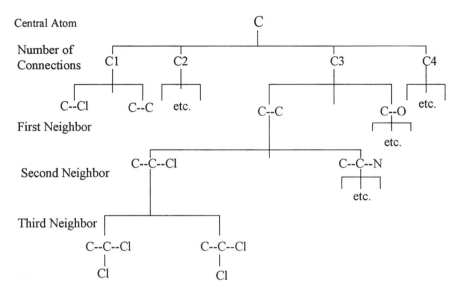

Figure 5.1. The hierarchical fragment descriptions used in the CIS (Feldmann) substructure search system. Each level in the hierarchy enlarges the description of the fragment.

level in the hierarchical fragment tree is effectively part of a hierarchical classification of all the atoms in the database as a whole, initially by number of neighbors, and atom type, and then by bonding pattern, and atom type of neighbors. The sizes of any rings in which the atoms occur is also taken into account, and lower levels of the tree are generated by applying a relaxation procedure, subdividing the atoms on the basis of the classifications already applied to their neighbors, and thus taking account of successively more distant atoms. Attached to each "leaf" at the bottom of the tree is a list of compound numbers for the compounds that contain the relevant atom type, and searching proceeds by tracing down the branches of the tree appropriate to the atom types (and their environments) in the query, and then merging the lists of compound numbers at the bottom.

This approach concentrates the computationally intensive work into creation of the hierarchical search files, leading to rapid searches and effectively obviating the need for an atom-by-atom search stage. However, for more general searches, where many branches of the tree must be traced and many lists merged, the system can be much slower. Initial creation of the tree search files is also very slow, though it is possible to update (rather than recreate) these files when new compounds are being added to the database. An advantage of the approach for large databases is that the size and complexity of the search files tends to "plateau off" as the database size increases. Thus the search times tend to be relatively better for large databases than for small ones; in contrast, the search time for conventional atom-by-atom matching is directly proportional to the number of compounds that pass the screening stage.

The S^4 system was developed by Softron GmbH in association with the Beilstein Institute, and it is used for substructure searching of the Beilstein database in the DIALOG Online system[30] and in the Beilstein Current Facts CD-ROM product.[31] In the S^4 system,[32] very compact codes are generated (again using a relaxation procedure) for each atom in the database, and they are sorted into a hierarchical search tree. The tree is arranged so that the minimum number of disk accesses is required during search, which further increases the speed, especially in CD-ROM implementations. Similar approaches are used in the ReSy system, developed internally for in-house use by Bayer AG at Leverkusen, Germany, and in the MACCS FastSearch system,[33] developed by MDL Information Systems, Inc., in the United States.[21]

The S^4 approach has been further refined in the CrossFire system, described in the remainder of this chapter, which has also been developed by Softron GmbH in association with the Beilstein Institute, and it is intended to operate in a client–server environment in in-house implementations.

The development of systems based on hierarchical tree searches during the past 10 to 15 years has represented a major advance in substructure searching software, and it now allows large databases to be searched in an acceptable amount of time on desktop microcomputers. The superior performance of tree-structured searching, in which much of the computational effort is put into preprocessing the database, compared to the conventional fragment screen and backtracking algorithm approach, was indicated by a comparison of systems carried out at the Beilstein Institute,[34] though the practical problems of adequately testing all the systems under identical conditions make it difficult to reach firm conclusions about the "best" system in all situations. Nevertheless, extremely efficient implementations of conventional fragment screening and backtracking atom-by-atom searches are also possible, and are used in recently developed systems such as CambridgeSoft's ChemFinder[35] and Oxford Molecular's RS3 Discovery system.[36] In the latter, the fragment-screening stage is performed within the standard Oracle database management system, although independent software is used for the atom-by-atom search stage.

The CrossFire Server—General Architecture

The architecture of the CrossFire Server is kept very simple. In fact, the online module consists of only a single program named XFIRE, which reads commands from standard input (the console) and writes the results as ASCII text to standard output (again, the console). However, nobody would ever use XFIRE by typing commands and interpreting the ASCII data stream. Intelligent graphical client software for PCs and Macintosh computers has been developed by Beilstein and Softron; this software hides all the tedious ASCII communication from the end user. The client opens a standard Telnet or Rexec session on the host machine and then starts XFIRE. Standard input and output of the host are then routed to the client, which has to interpret the ASCII stream. By architecture, XFIRE is also scriptable. On Unix machines, for example, one could use the standard pipe mechanism to feed XFIRE with commands and direct the output to some program for analysis. Starting the XFIRE program establishes a peer-to-peer connection between a single end user and the CrossFire host. During an open session, XFIRE maintains an automatic history of all temporary hit sets. When the user logs out, the history is cleared. Only hit sets that are saved explicitly by the user are available for future sessions. This chapter will discuss some of the kernel algorithms of CrossFire.

The CrossFire Structure Retrieval Engine

The CrossFire structure search engine has been developed and is still being improved by Softron GmbH in close cooperation with Beilstein Information Systems GmbH. A lot of key know-how in the field of chemical structure databases was acquired by Softron when building the S^4 search engine[32] for the implementation of the Beilstein File on the DIALOG Online system[30] and the Beilstein Current Facts edition on CD-ROM.[31]

The CrossFire structure search engine implements a two-stage method consisting of a screening filtering step and a final atom-by-atom search (ABAS). Although other approaches to substructure searching tried to eliminate the atom-by-atom step in favor of a more comprehensive screening, they could never prove in a theoretical mathematical sense that the search results would always be correct. In practice, the results seemed to be correct for the sets of chemical structures tested, but it is always possible to find (theoretical) counterexamples. Therefore, to guarantee ultimate reliability in the CrossFire system, we decided to let all hit structures pass an ABAS.

When the CrossFire search engine was designed, the main intention was to develop a fast retrieval engine for the Beilstein File. This file is very large (7.3 million compounds at present) and is updated every 3–4 months in portions of several tens of thousands of structures. Therefore, dynamic update of the database was not a primary design goal, but we wanted to concentrate on fast retrieval. We had learned from the analysis of many other substructure search systems that queries usually execute faster if only a few disk seeks are required. Traditional inverted list methods using fixed or automatic screen libraries[2,3,34,37] always produce sets of primary keys [e.g., registry numbers (RNs)] after the screening step. Since the structure records are distributed randomly in the file, the ABAS usually requires one disk seek per record to load it into memory for the final match. This is acceptable for small candidate sets but becomes impracticable for unspecified queries producing large candidate sets. Traditional inverted list methods are input–output bound. However, the number of disk seeks could be reduced dramatically if the candidates found by the screening were always in sequential order. Of course, this requires some preprocessing and prearranging of the original structures. The basics of structure coding and searching in CrossFire are described below.

Structure Coding

The basis of structure coding in CrossFire is a unique numbering of all atoms in the molecule. With substructure searching in mind, we construct a unique numbering starting with a fixed center atom. Based on atom and bond characteristics, the numbering chosen in the CrossFire system has the following properties:

1. the center atom has index 1,
2. all atoms of the same sphere span a contiguous interval of indexes,
3. all atoms with same root atom span a contiguous interval of indexes, and
4. atom X has a lower index than Y if X has a lower indexed root atom (*see* Figure 5.2).

Atoms with the same root atom are grouped together to form so called "bundles", and the center atom itself is defined as bundle 0. Note that in many substructures you still find complete bundles, whereas complete spheres are rare because of free sites or undefined bonds.

This numbering is used to produce a series of linear representations ("bit strings") for each structure graph by choosing each graph atom as the center in turn. Note that the numerical representation of the center atom is independent of its location within the structure: a carbon atom with two neighbor atoms has the same numerical value, no matter what the neighbors are. More generally, this is also true for whole bundles. In other words, there is a 1 to 1 correspondence between identical (partial) bit strings and identical (sub)structures.

All bit strings of all structures are sorted into the so-called CrossFire Search File. Each bit string carries the registry number of the structure it represents. Taking a closer look at the Search File, we find that it is clustered in a natural way. All structures containing a certain atom, such as a sulfur atom with two neighbors, appear in sequence in the Search File. All sulfurs with the *same* neighbors also appear in sequence, and so on. And the latter sequence is always completely included in the previous one.

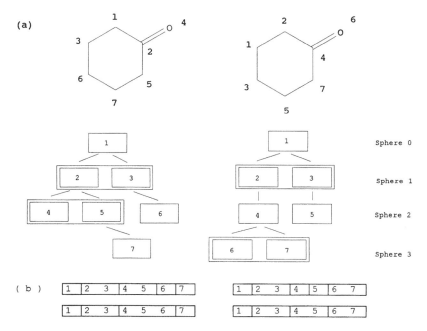

Figure 5.2. (a) Two numberings of cyclohexanone based on different center atoms. (b) Different bit strings resulting from the numberings, grouped by bundles (upper) and spheres (lower), respectively.

Index Files

For the screening, we built special index files (the "bundle files"). All bit strings in the Search File are split on bundle boundaries and reorganized in a single rooted tree. The bundles for different layers are stored in different physical files. The first CrossFire version used 11 layers (0 . . . 10) which gave good results for most queries. But after some time we found that dodecane, which is a simple chain of 12 saturated carbon atoms connected with single bonds (a full structure!) led to unexpectedly long search times. In contrast, undecane, a chain of 11 carbon atoms, was found quickly.

We analyzed the problem and found the following. The screening had worked out all 11 bundle levels but, of course, it had to leave one carbon atom unmapped. The interval defined by this bit string, however, was very large: it contained all chains of 12 *and more* carbons, i.e., more than 80,000. If we had used 12 bundle levels instead of 11, everything would have worked fine—until we asked for a chain of 13 carbon atoms. A simple solution to this problem seemed to be to use more bundle files, say 20 instead of 11. We tried that—and nearly ended up with an out-of-space error from the file system. The reason: bundle files tend to increase in size dramatically. The solution was to build 20 bundle files but not to store all 20 bundles of each bit string. Instead, we stop storing bundles of a bit string at level K, if the *partial* bit string up to level K is already unique in the whole file! This is often true for bit strings with one or more selective atoms in the first layers. Many structures share only the first 7 to 10 bundles with others. With this technique we achieved an overall size of 20 new bundle files no larger than the 11 bundle files before. And retrieving dodecane was no longer a problem.

Structure Search

The screening is a process of iterative key refinement. First, using a heuristic algorithm, the "best" starting atom in terms of the most selective concentric environment is fixed in the query. All 0-bundles matching the starting atom are picked from Bundle File 0. Note that Bundle File 0 contains all different atoms of the structure file.

Let us concentrate on one single 0-bundle. It is the key to a section in the Search File containing candidates for the ABAS. This section is very large if many bit strings start with this 0-bundle. A 0-bundle is the root of a subtree in Bundle Files. By branching from the 0-bundle, we find all feasible 1-bundles. Only those 1-bundles are selected that match the corresponding atoms in the query. Note that by construction of the bit strings, it is always known which atoms of the query correspond to the actual bundle. A matching 1-bundle is concatenated to the 0-bundle, thus yielding a more specific key to the Search File. This procedure is iterated until the last Bundle File is reached or until all atoms of the query have been mapped. Each iteration reduces the number of candidates for the ABAS.

The CrossFire screening uses ABAS-like algorithms, and since it has the complete connection table information up to the current bundle level available, it can decide when it was exact for a query, thus making an ABAS match unnecessary. This means an enormous speed-up of searching, especially for small unspecified queries.

Efficiency Control

We have said that sometimes searches took a long time because the screening used too many bundle levels. The overall search time could have been shortened if the screening would have stopped before mapping all atoms of the query. The reason is that the ABAS is much more efficient in dropping false hits than the screening can be. So we looked for a decision strategy to optimize the overall system performance by stopping the screening before it becomes too inefficient. On the other hand, of course, a higher atom-by-atom load must be taken into account. We took the following approach.

Let W be the overall CrossFire "work" (in an abstract unit, for now) for a single query. W shall be minimized:

$$W \to \min$$

In the CrossFire system, W is the sum of W_S and W_A, the work done by the screening and the work done by the atom-by-atom load, respectively.

$$W = W_S + W_A \to \min$$

The work depends on the number, L, of bundle levels used:

$$W(L) = W_S(L) + W_A(L) \to \min$$

The screening work is the sum over all works per bundle level:

$$W_S(L) = w_S(0) + \cdots + w_S(L)$$

It can be seen from the formula that the screening work is a monotonically increasing function of the bundle level, L. On the other hand, one can expect the ABAS work to be a decreasing function of L. The question is, how shall L be selected to minimize the overall work, W? To answer this question, we claim that $W(L)$ monotonically decreases with L:

$$W(L + 1) \le W(L), L = 0, 1, 2, \ldots$$

It follows that

$$w_S(0) + \cdots + w_S(L) + w_S(L + 1) + W_A(L + 1) \le w_S(0) + \cdots + w_S(L) + W_A(L)$$

which is equivalent to

$$w_S(L + 1) + W_A(L + 1) \le W_A(L)$$

Let us make the following assumptions for $W_A(L)$:

1. The work for just reading RNs from the Search File without an actual match can be neglected.

2. The work of $W_A(L)$ is determined by the number of structures, $K(L)$, that have to pass an actual match:

$$W_A(L) = W_A[K(L)]$$

3. The number of candidates, $K(L)$, monotonically decreases with L, and $K(L)$ can be expressed as the sum of candidates after screening step

$L + 1$ plus the number of candidates dropped by the screening in step $L + 1$, $D(L + 1)$:

$K(L + 1) \leq K(L)$ (K monotonically decreasing)

$K(L) = K(L + 1) + D(L + 1)$

Together, we get

$w_S(L + 1) + W_A[K(L + 1)] \leq W_A[K(L)] = W_A[K(L + 1) + D(L + 1)]$

$w_S(L + 1) + W_A[K(L + 1)] \leq W_A[K(L + 1)] + W_A[D(L + 1)]$

$w_S(L + 1) \leq W_A[D(L + 1)]$, $L = 0, 1, 2, \ldots$

$w_S(L) \leq W_A[D(L)]$, $L = 1, 2, \ldots$ **(Breakoff criterion)**

That means that the screening work in step L (i.e., dropping $K(L)$ candidates) is less than or equal to the respective work done by the ABAS (i.e., processing the $K(L)$ candidates). Although $W_A[D(L)]$ is unknown during the screening, an estimation value, $W_{A[est]}$, can be computed instead. Thus, the screening must (1) count the number of candidates, $D(L)$, dropped in each step, (2) measure the work, $w_S(L)$, for each screening step, and (3) compare $w_S(L)$ with $W_{A[est]}$. If $w_S(L) > W_{A[est]}$ for some L, the screening is stopped. With the breakoff criterion enabled, we found a remarkable speedup, especially for very unspecified queries, such as with "free sites" on all atoms. For an example, *see* Figure 5.3.

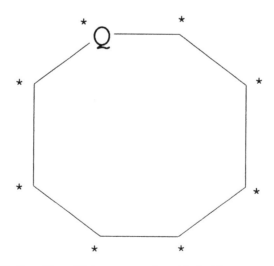

Figure 5.3. With CrossFire efficiency control enabled, this sample query uses only 6 instead of 20 bundle levels, and the overall execution time is 25% faster.

Reaction Retrieval

The CrossFire retrieval engine for chemical reactions is designed as a straightforward extension of the structure retrieval engine (*see* the section "The CrossFire Structure Retrieval Engine"). Reactions are conceptually treated as multicomponent structures. Thus, we were able to reuse a large portion of the approved retrieval code. We added some component attributes to mark educt reactants and product reactants, flagged reaction center atoms, and inserted atom-mapping information in a compressed format into each bit string. The CrossFire Search File and the Bundle Files are analogous to structure searching. Reaction searching is then performed in the following steps. The screening scans through the Reaction Bundle Files and provides the ABAS with partial bit strings. Only structural attributes are checked by the screening, and all restrictions requiring the atom-mapping (such as broken bonds, a change in bond type, the atom-mapping itself, etc.) are not covered. The ABAS performs the final structure match and verifies all reaction attributes requested by the query.

User Databases

Although the CrossFire search engine is designed and tuned for retrieval, it is possible to store and retrieve individual chemical structures interactively. In order to keep search times fast even for large user databases, CrossFire implements a logical database as one enterprise partition (EP) and several department partitions (DPs). The EP of the database uses the fast CrossFire index files for retrieval, while the DPs are only searched sequentially. The DPs are meant for interactive update, whereas the EP is read-only. Each DP "owns" a nonoverlapping region of the whole user database and is coupled with a unique user ID. Or, the other way around, each user has control over a certain region of the database while the remainder, that is, the EP, remains static from the user's point of view. Within the user's region, new structures can be entered, deleted, or updated. These changes are immediately reflected in the user's own search results, but they have no effect on searches of all other users.

In order to synchronize the whole database, the system administrator must use the CrossFire maintenance program to merge all DPs into the EP and to index it with the CrossFire file loading programs. After the merge, all changes to the DPs since the last synchronization are now visible to all users. Moving structures from the DPs to the EP does not affect the ownership of the structures. The owner can still edit or delete them and will see the changes immediately.

The advantage of using this database concept is that structure searches can benefit from the fast CrossFire retrieval algorithm and that search results can be consistently stored in temporary files. On the other hand, concurrent update of the database by all users simultaneously is not possible.

In the current CrossFire Version 3.2, the user databases can only contain records consisting of one primary key (as a 32-bit integer number) and a chemical structure connection table (as a variable-length text string). Additional keys and data are planned for the future.

References

1. Tarjan, R. E. "Graph Algorithms in Chemical Computation," in *Algorithms for Chemical Computations*, Christoffersen, R. E., Ed. *ACS Symposium Series*; American Chemical Society, Washington, DC, 1977, *46*, pp 1–19.
2. Willett, P. "A Review of Chemical Structure Retrieval Systems," *J. Chemometrics* **1987**, *1*, pp 139–155.
3. Barnard, J. M. "Substructure Search Methods: Old and New," *J. Chem. Inf. Comput. Sci.* **1993**, *33*, pp 532–538.
4. Karp, R. M. "On the Computational Complexity of Combinatorial Problems," *Networks* **1975**, *5*, pp 45–68.
5. Farmer, N.; Amoss, J.; Farel, W.; Fehribach, J.; Zeidner, C. R. "The Evolution of the CAS Parallel Structure Searching Architecture," in *Chemical Structures: The International Language of Chemistry (Proceedings of an International Conference at the Leeuwenhorst Congress Center, Noordwijkerhout, Netherlands, 31 May–4 June, 1987)*; Warr, W. A., Ed.; Springer: Heidelberg, 1988; pp 283–296.
6. Jochum, P.; Worbs, T. "A Multiprocessor Architecture for Substructure Search," in *Chemical Structures: The International Language of Chemistry (Proceedings of an International Conference at the Leeuwenhorst Congress Center, Noordwijkerhout, Netherlands, 31 May–4 June 1987)*; Warr, W. A., Ed.: Springer: Heidelberg, 1988; pp 279–282.
7. Downs, G. M.; Lynch, M. F.; Willett, P.; Manson, G. A.; Wilson, G. A. "Transputer Implementations of Chemical Substructure Searching Algorithms," *Tetrahedr. Comput. Methodol.* **1988**, *1*, pp 207–217.
8. Wipke, W. T.; Rogers, D. "Rapid Subgraph Search Using Parallelism," *J. Chem. Inf. Comput. Sci.* **1984**, *24*, pp 255–262
9. Willett, P.; Wilson, T.; Reddaway, S. F. "Atom-by-Atom Searching Using Massive Parallelism. Implementation of the Ullmann Subgraph Isomorphism Algorithm on the Distributed Array Processor," *J. Chem. Inf. Comput. Sci.* **1991**, *31*, pp 225–233.
10. Ray, L. C.; Kirsch, R. A. "Finding Chemical Records by Digital Computers," *Science* **1957**, *126*, pp 814–819.
11. Dengler, A.; Ugi, I. A. "Central Atom Based Algorithm and Computer Program for Substructure Search," *Comput. Chem.* **1991**, *15*, pp 103–107.
12. Xu, J. "GMA: A Generic Match Algorithm for Structural Homomorphism, Isomorphism, and Maximal Common Substructure Match and Its Applications," *J. Chem. Inf. Comput. Sci.* **1996**, *36*, pp 25–34.
13. Morgan, H. L. "The Generation of a Unique Machine Description for Chemical Structures—A Technique Developed at Chemical Abstracts Service," *J. Chem. Doc.* **1965**, *5*, pp 107–113.
14. Sussenguth, E. H. "A Graph-Theoretic Algorithm for Matching Chemical Structures," *J. Chem. Doc.* **1965**, *5*, pp 36–43.

15. Figueras, J. "Substructure Search by Set Reduction," *J. Chem. Doc.* **1972**, *12*, pp 237–244.

16. Ullmann, J. R. "An Algorithm for Subgraph Isomorphism," *J. Assoc. Comput. Mach.* **1976**, *23*, pp 31–42.

17. von Scholley, A. A. "Relaxation Algorithm for Generic Chemical Structure Screening," *J. Chem. Inf. Comput. Sci.* **1984**, *24*, pp 235–241.

18. Brint, A. T.; Willett, P. "Pharmacophoric Pattern Matching in Files of 3D Chemical Structures: Comparison of Geometric Searching Algorithms," *J. Mol. Graphics* **1987**, *5*, pp 49–56.

19. STN International, c/o Chemical Abstracts Service, 2540 Olentangy River Road, P.O. Box 3012, Columbus, OH 43210–0012.

20. Questel Orbit Groupe France Telecom, Le Capitole, 55 avenue des Champs Pierreux, 92029 Nanterre Cedex, France.

21. MDL Information Systems, Inc., 14600 Catalina Street, San Leandro, CA 94577. Accessible via the WWW at URL: http://www.mdli.com.

22. Hampden Data Services Ltd., Kingmaker House, Station Road, New Barnet, Herts EN5 1NZ, U.K.

23. Berks, A. H. "The Chapman and Hall Dictionary of Drugs on CD-ROM," *J. Chem. Inf. Comput. Sci.*, **1995**, *35*, pp 332–333.

24. Dittmar, P. G.; Farmer, N. A.; Fisanick, W.; Haines, R. C.; Mockus, J. "The CAS Online Search System. 1. General System Design and Selection, Generation and Use of Search Screens," *J. Chem. Inf. Comput. Sci.* **1983**, *23*, pp 93–102.

25. Daylight Chemical Information Systems, Inc., 27401 Los Altos, Suite #370, Mission Viejo, CA 92691. http://www.daylight.com.

26. Chemical Design Ltd., Roundway House, Cromwell Park, Chipping Norton, Oxfordshire OX7 5SR, U.K.

27. Feldmann, R. J.; Milne, G. W. A.; Heller, S. R.; Fein, A.; Miller, J. A.; Koch, B. "An Interactive Substructure Search System," *J. Chem. Inf. Comput. Sci.* **1977**, *17*, pp 157–163.

28. Attias, R. "DARC Substructure Search System: A New Approach to Chemical Information," *J. Chem. Inf. Comput. Sci.* **1983**, *23*, pp 102–108.

29. Nagy, M. Z.; Kozics, S.; Veszpremi, T.; Bruck, P. "Substructure Search on Very Large Files Using Tree-Structured Databases," in *Chemical Structures: The International Language of Chemistry (Proceedings of an International Conference at the Leeuwenhorst Congress Center, Noordwijkerhout, Netherlands, 31 May–4 June 1987)*, Warr, W. A., Ed.; Springer: Heidelberg, 1988; pp 127–130.

30. Knight–Ridder Information, Inc., 2240 El Camino Real, Mountain View, CA 94040. http://www.krinfo.com.

31. Hicks, M. G. "Beilstein Current Facts in Chemistry: A Large Chemical Database on CD-ROM," *Anal. Chim. Acta* **1992**, *265*(2), pp 291–300.

32. Bartmann, A.; Maier, H.; Roth, B.; Walkowiak, D. "Substructure Search on Very Large Files by Using Multiple Storage Techniques," *J. Chem. Inf. Comput. Sci.* **1993**, *33*, pp 539–541.

33. Christie, B. D.; Leland, B. A.; Nourse, J. G. "Structure Searching in Chemical Databases by Direct Lookup Methods," *J. Chem. Inf. Comput. Sci.* **1993**, *33*, pp 545–547.

34. Hicks, M. G.; Jochum, C. "Substructure Search Systems. 1. Performance Comparison of the MACCS, DARC, HTSS, CAS Registry MVSSS, and S^4 Substructure Search Systems," *J. Chem. Inf. Comput. Sci.* **1990**, *30*, pp 191–199.

35. Cambridgesoft Corporation, 875 Massachusetts Avenue, Cambridge, MA 02139. http://www.camsci.com.
36. Oxford Molecular Ltd., The Medawar Center, Oxford Science Park, Oxford OX4 4GA, U.K. http://www.oxmol.co.uk.
37. Barnard, J. M. "Problems of Substructure Searching and Their Solution," in *Chemical Structures*; Warr, W. A., Ed.; Springer–Verlag: Berlin, 1988; pp 113–126.

Chapter 6

CrossFire*plus*Reactions

Alexander J. Lawson

The design concept and history of CrossFireplusReactions are discussed in detail from the standpoint of the underlying database and architecture, the performance of the search and retrieval engines, and the unifying client–server interface. The principles are illustrated by three worked examples in the area of information retrieval for organic reactions.

T he CrossFire concept was based on the postulate that any information system in chemistry must *simultaneously* satisfy three criteria to be effective:

1. the underlying database must be extensive in coverage and detailed in its indexing,
2. the search and retrieval systems must be powerful, particularly for chemical subgraphs, and
3. the user interface must reflect the user's needs in his or her own context.

This chapter will discuss each of these three aspects in turn. But before proceeding, it is worthwhile noting that before CrossFire, computerized information on chemistry largely meant CAS on STN or DIALOG in the minds of many users, for the following reasons:

1. Users demanded and needed large files of information.
2. Therefore, they demanded a centralized computer center containing expensive main frames and disk drives.
3. The user interface and searching capability were then particularly difficult to use, because the command line interfaces of main frame computers are notoriously cryptic.

The advent of CrossFire has therefore widened the choice. CrossFire is a revolutionary concept in that it breaks the chain of this logical dependency at the second step (2): the files involved in the CrossFire solution are indeed huge, but the search algorithms are fast enough to make the

© 1998 American Chemical Society

2,3,5-Triphenyl-2,3-dihydro-tetrazol-1-yl $C_{19}H_{15}N_4$, formula IV.
 Prep. From 2,3,5-triphenyl-tetrazolium chloride, by treatment with Ag-Hg (*Y. Deguchi, Y. Takagi*, Tetrahedron Lett. **1967** 3179), or with tetra-*p*-tolyl-hydrazine [benzene; N_2; 60°] (*F. A. Neugebauer*, Tetrahedron Lett. **1968** 2129; *F.A. Neugebauer, C.A. Russell*, J. Org. Chem. **33** [1968] 2744). — ESR spectrum and hyperfine coupling constants (*Ne.*).

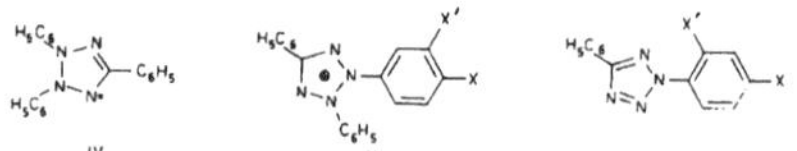

2-(4-Fluoro-phenyl)-3,5-diphenyl-tetrazolium $[C_{19}H_{14}FN_4]^+$, formula V (X = F, X' = H).
 Perchlorate. Prep. From 1-(4-fluoro-phenyl)-3,5-diphenyl-formazan, by treatment with (i) $Pb(OAc)_4$ [AcOH], and (ii) aq. $KClO_4$ (*G. Arnold, C. Schiele*, Spectrochim. Acta Part A **25** [1969] 671, 672). — Cryst. [from MeOH]; IR spectrum.
 2-(3-Chloro-phenyl)-3,5-diphenyl-tetrazolium $[C_{19}H_{14}ClN_4]^+$. Perchlorate. Cryst. [from MeOH]; IR spectrum (*Ar., Sch.*).

2-(4-Chloro-phenyl)-5-phenyl-2H-tetrazole $C_{13}H_9ClN_4$, formula VI (X = Cl, X' = H).
 Prep. From benzaldehyde (4-chloro-phenyl)-hydrazone and PhN_3 [Na; 2-methoxy-ethanol; 110 – 115°] (*S.-Y. Hong, J.E. Baldwin*, Tetrahedron **24** [1968] 3787, 3793). From *N'*-benzenesul*-fonyl-benzohydrazonoyl chloride and (4-chloro-phenyl)-hydrazine [O_2; K_2CO_3; THF] (*S. Ito et al.*, Bull. Chem. Soc. Jpn. **49** [1976] 762, 763). From benzaldehyde benzenesulfonylhydrazone and 4-chloro-benzenediazonium chloride [Py] (*S. Ito et al.*, Bull. Chem. Soc. Jpn. **49** [1976] 1920, 1922). — mp: 123 – 124° [from EtOH] (*Ito et al.* 1920). Brown cryst.; mp: 120.5 – 121.5°; UV (*Hong, Ba.* 3788). IR; UV (*Ito et al.* 764). — Thermal decomposition [1-chloro-naphthalene; 165.8°]: kinetics (*Hong, Ba.* 3789).

2-(4-Chloro-phenyl)-3,5-diphenyl-tetrazolium $[C_{19}H_{14}ClN_4]^+$, formula V (X = Cl, X' = H).
 Chloride. IR spectrum; UV (*T. Pukas*, Zesz. Nauk. Politech. Slask. Chem. no. 15 [1963] 1, 29, 67; CA **62** [1965] 8644).

Figure 6.1. Typical section of the *Beilstein Handbook*.

use of a main frame computer unnecessary. (*See* Chapter 5 for more details on this aspect of CrossFire.) Thus, the danger of a quasi-monoculture in chemical information was avoided, to the benefit of the user community at large, but also to the benefit of all database producers, information specialists, and online hosts, who were often unhappy (even to the point of litigation) with the dependencies involved in the developing mono-culture. However, the major benefits of CrossFire were realized by end-user chemists, for whom the system was principally designed. It is interesting to now trace the development of CrossFire in the order of the above points (1–3).

The Beilstein File

The inherent potential of Beilstein as an information source has always been clear upon opening any page of the well-known *Beilstein Handbook*, as seen in Figure 6.1. The essential features of the Beilstein "world view" are clear:

- The work is organized as a list of organic structures, and indeed it has always been accessible directly by structural algorithm, long before computers were conceived (the "Beilstein System" of organization is an access tool that requires only structural elements, and it goes back in principle to Konrad Beilstein himself).
- For each structure, there are compiled individual data elements or fields (of data or information), each with its own set of literature citations.

- The data elements are basically of two types: synthesis-related (e.g., data on preparation and chemical behavior), and property-related (e.g., physicochemical numerical data, such as viscosities, characterization data, pKs, etc.).
- Closer inspection of the printed page will illustrate that ca. 50% of the data elements are concerned with the synthetic part, and further investigation will reveal that essentially all of the starting materials cited in the text do themselves possess a Beilstein entry at another point in the Handbook, in which their preparations are detailed in turn.

The conversion of the rich Beilstein data source into electronic form and its availability on STN and DIALOG from 1989 should have opened up, in principle, the full potential of this fascinating work as a multidimensional source of information from four complementary and interactive standpoints: structures, reaction steps, property data, and citations (documents).

In retrospect, it appears that this was only partially achieved. The data structure as implemented on these online hosts shows a heavy emphasis toward chemical structure searching and display for individual entries and their numerical properties. However, the advantages of searching for structural information in a set of reaction steps (i.e., multiple structures) on a structural (connection table) basis and the data elements of individual citations (documents) were not exploited and made available at that time. Mentioning this deficit here is by no means intended as a criticism of the database or online implementations; rather, the deficit is a direct consequence of two major inherent characteristics of online chemical services: information specialists and connect-hour pricing. The link between these apparently unrelated aspects can be shown by the following argument:

- Online services have always had (until very recently) a pricing system that is heavily linked to connect-hour charges (i.e., how long a session lasts, also called the "taxi-meter syndrome").
- Online services have a complex and even nonintuitive query language. It is also true that this query language tends to be kept basically invariant because it serves many databases on the same host, and users (once they have mastered the syntax) will be resistant to change. Thus, individual newcomer databases must fit the existing house rules if they are to find widespread use.
- Online services are generally centralized, and they use search engines that cater to the whole world simultaneously. They are therefore not quick enough in their responses (particularly in substructure searching) to be regarded as interactive. Such searches are expensive.
- It is therefore desirable to prepare a strategy beforehand to keep costs down (by avoiding browsing).

The result of all these factors is that an expert who is familiar with the system (an information specialist, or intermediary) will most likely become involved or even take over the whole search contract. Involvement of an expert leads to the following:

- Any chance of spontaneous creativity is practically annulled (because only the end user can work creatively with his problem on the fly).
- Switching contexts into a document-based view and/or following reaction paths at will are both cases in which the user is looking into areas in hopes of finding something interesting, whereas structures and properties are hard-and-fast data which are either present or not. Information specialists are understandably less prone to follow a speculative path on behalf of their clients, where it is not clear to them that they are on an intrinsically interesting path; they are generally less inclined to spend money on potentially unwanted information and would rather seek concrete, hard-and-fast data.

Given that this is an oversimplification of the complex issues involved in retrieving information in the 1980s and early 1990s, it is nevertheless understandable why the implementation of the Beilstein file at STN and DIALOG followed the course it did, and why the inherent potential of the file was only partially realized. This fact was recognized by Beilstein as early as 1990, and plans to develop a long-term alternative access mode to the Beilstein file were initiated in 1991. At that time the question of hardware price/performance ratios (particularly in the case of disk storage) seemed to preclude a decentralized solution, and the phrase "client–server" was certainly not in common usage. Nevertheless, Beilstein interpreted the progress curves of the hardware industry as indicating that the solution they had in mind would be technically meaningful by the time it had been fully developed and tested, and work was started. Key input over the two-year initial development period was provided by a team of consultants from the chemical industry, the computer industry, and academe. Between 1993 and 1995 the CrossFire system rapidly evolved from a structures-only module, through structures and properties (thus emulating the Beilstein file on STN and DIALOG), to the present solution, which incorporates chemical reactions in a natural graphic form.

The first important insight was the viability of the client–server concept (Figure 6.2). This concept involves an architecture in which the file is maintained on mass storage at a workstation server equipped with sufficient disk capacity and computing power for searches to be carried out in a reasonable response time. The intrinsic desirability of such an arrangement was stressed from the very start by all industrial advisers, because it opens the way to creative interaction of the end user without necessarily involving either intermediaries or connect-hour costs. Knowing in advance that the costs for all searching in a year would be fixed or by subscription was considered a very attractive feature of this approach.

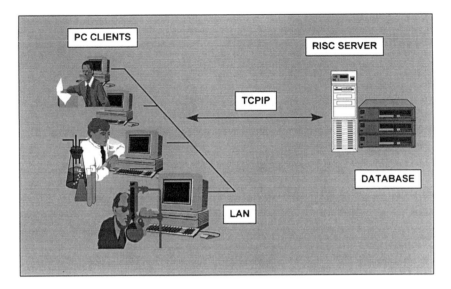

Figure 6.2. CrossFire client–server architecture.

Thus, an annual subscription fee replaced pay-as-you-go, and nearly every potential subscriber in the market regarded this as a decisive bonus from the point of view of work efficiency and budget planning.

The essential precondition at this point was (and is) performance, rather than the more obvious caveat of hardware costs; most strategic managers could accept the notions that hardware price/performance ratios would continue to improve, that operating system software integration would become less of a headache, and that graphical user interfaces would converge on some de facto standards. But the key aspect remained refreshingly focused for the Beilstein development team: could they achieve search performance with the workstation server that would be much better than the main-frame-based alternative online host? This became the heart of the attempt to finally open the gate to truly interactive use of their huge file; all other aspects were doubtless also important but were regarded more as "further details".

The Search and Retrieval Performance of CrossFire

Structure Searching

The first indication of an answer to the question posed in the preceding section was demonstrated in 1993, when a benchmark study[1] of the CrossFire system for approximately 5 million structures found that the substructure searching system was between 1 and 2 orders of magnitude quicker in

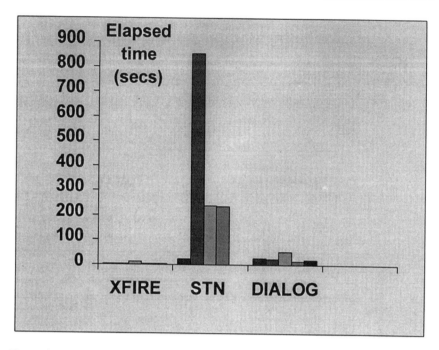

Figure 6.3. Response times for search and retrieval of five typical substructure queries in 1993.

practice than systems using the corresponding structure files on online hosts. (*See* Chapter 5 for details on how CrossFire substructure searching works.) The graphic representation of the elapsed time measured in that study is reproduced here in Figure 6.3: it can be seen that the response times of CrossFire are barely detectable in comparison to the results obtained using DIALOG or, especially, STN.

Text and Numerical Searching

Similarly, nongraphical elements such as text, numerical, and keyword searching and the manipulation of hit sets (in particular, their Boolean intersection) must also be very fast. This is a feature of CrossFire that has not been widely investigated, but the figures are impressive. Without going into detail, the following two facts are worth considering:

- In pure text searching, CrossFire supports not only right truncation, but also left and middle truncation in any combination. For example, the word "antileukemic" occurs frequently in the original literature in the context of biological testing. It is a moot point whether this is a proper word, but the original authors used it as such, and therefore

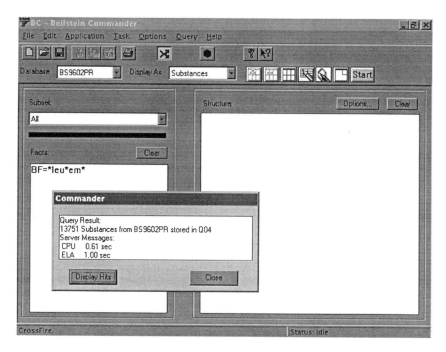

Figure 6.4. Response time for a truncated text query in the context of Biological Functionality.

we have to accept that fact. But such words are prone to a variety of spelling variations, such as in the use of "leukaem . . ." or even "leucemic". CrossFire automatically links the biological activity to the context of the compound concerned (a great advantage over purely bibliographic databases), but to ensure good coverage of the subcontext with respect to this illness, it is advisable to use truncation, such as "*leu*em*". This use of wild cards at the left and in the middle is very seldom met with in large text databases, because the search engines are not capable of handling such a vague description. Figures 6.4 and 6.5 show how quickly such a query was handled by CrossFire when searching ca. 7 million substances. Naturally, the wild card "?" is also supported (for one character), in keeping with DOS conventions.

- On the related issue of the intersection of the lists, CrossFire knows no system limits. No hit set is too large to be handled. For example, intersection of the two lists for each compound with a melting point (more than 3 million) and each compound with a boiling point (more than 0.6 million) took less than 4 s when an IBM RS/6000 workstation was used as a server. This is a chemically obscure use of list intersection, but the size of the lists makes the point concerning system limits and speed of performance.

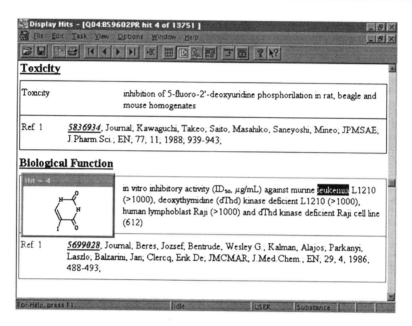

Figure 6.5. Display of one of the hits from the query of Figure 6.4.

The User Interface

CrossFire runs on both IBM/Windows and Apple Macintosh client platforms. The software is essentially identical in its look and feel, although minor differences exist because of the intrinsic differences in the platform philosophies of these two major systems. As in practically every system, there is a query level and a display level. The content of the answer sets is always a mix of (structural) graphics and text, and these are integrated for display purposes. On the query level, these are handled by separate modules, and previously obtained hit sets can also be incorporated into the query with Boolean logic. The true cockpit of the session, however, is the context controller, called the "Commander".

This arrangement is shown in schematic form in Figure 6.6 and with screen shots in Figure 6.7. The philosophy behind this arrangement is simple enough: a topic to be searched is couched in terms of a combination of numerical, textual, and graphical objects (using wild cards, with range-searching and partial-structure attributes where necessary) and combined as required with any hit set of any of the three types of database objects [substances, reactions, and documents (or "citations" in Beilstein terminology)].

This is the combined work of the Commander (context level) and the two query editors underlying the Commander. On starting the search,

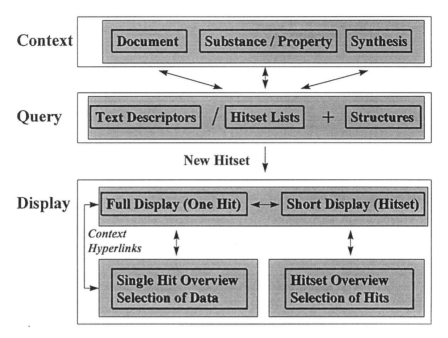

Figure 6.6. Schematic representation of the relationships between Context, Query, and Display.

this query is communicated via an ASCII protocol to the server wherever it sits (this could be literally anywhere, as tests using the Internet have shown). The server then carries out the search and returns the hit set in ASCII form to the client, where it is interpreted on-the-fly for display purposes.

It has long been an unofficial dogma at Beilstein that the subjective value of working with a system is directly proportional to the ratio of the time spent by the user actively working in the display mode (i.e., using the results) with respect to the total time taken up with the session. This is perhaps not strictly a mathematical theorem, but in practice it is a very good measure of user satisfaction. The display mode is accordingly equipped with a number of tools that enable the user to view and select a subset of the hit set objects, concentrate on these alone, view them singly, and, finally, reduce the view of these objects to the particular facets of interest. This process can be called "concentration of relevance", and it is one aspect of user requirements. However, an equally important functionality is to be able to *widen* the relevance at will, browsing into related aspects without necessarily having to reformulate a query (although this can always be carried out, of course). The major tool for widening the relevance is the *hyperlink*, the use of which is described in other chapters.

Context

Query

Display

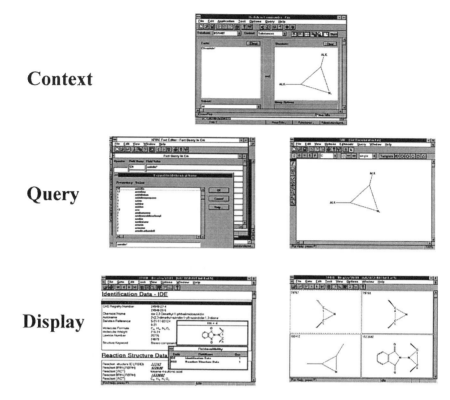

Figure 6.7. Screen shots of the modules corresponding to the concepts of Figure 6.6.

Any user of the World Wide Web (WWW) is familiar with the hyperlink concept, and it need not be explained in detail here. Attention should perhaps merely be drawn to the fact that the Beilstein file under CrossFire contains more than 30 million of these links, which puts the system at least into the same category as the WWW itself (at the time of writing, AltaVista[2] claims to index some 50 million pages of the WWW via hyperlinks.)

To summarize, by building on the fundamental power of the Beilstein search engines, hypertext linking was added on the client side in 1994–1995, reactions were added in 1995–1996, and document-based indexing was added in 1996–1997. These enhancements have made the CrossFire solution into a tool that fully exploits the underlying Beilstein data.

The CrossFire solution today treats the Beilstein file as a unit of three databases, tightly integrated in such a way that the user should never need to consider this underlying structure:

- a database of structures and properties,
- a database of reactions, and
- a database of documents.

Three other chapters in this book (Chapters 7–9) illustrate this in practice, concentrating on the Beilstein file under CrossFire. However, the reader should not lose sight of the fact that other files may be accommodated very comfortably under CrossFire, including company internal private files and the files from other vendors (such as the Gmelin inorganic file, which has been available under CrossFire since 1996).

One final point concerns user dynamics. The CrossFire user generally has a goal, whether great or small, general or specific, of finding out what the chemical literature says about a particular topic. In other words, the starting point of a session is the definition of the topic (which may change in the course of the session because of spontaneous creativity), and the final result is a pointer to the content of a relevant set of documents dealing with the required subject matter. Although in certain cases the actual content of the database may represent the end result itself, it is more usual for the user to consult the original documents for further detail. This last step is very time consuming, and therefore the quality of a database such as Beilstein lies in the quality of its indexing power, that is, its capability of reducing the suggested list of articles to a manageable size without discarding relevance. This is a task that only the end user can perform in the last analysis, and CrossFire pursues a policy of casting as wide a net as possible, and providing the user with tools to reduce the hit set to as relevant a set as possible in as short a time as possible.

This section has described the general CrossFire concept in some detail. We will now turn to CrossFire's handling of reactions.

Reactions under CrossFire

As mentioned above, CrossFire sees the world of organic chemistry as being largely defined by three types of objects: structures (and their properties), reactions (and their conditions), and documents (and their contents). Every topic can be fitted to a combination of these objects in this broad classification scheme, and it should also be noted that in the Beilstein approach each of these three units is mutually embedded in each of the other two, that is:

- Details linked to a structure include details of its preparation (which is a reaction) as well as details of studies of its physical properties, and each of the individual reports is linked to a literature reference to the document involved.

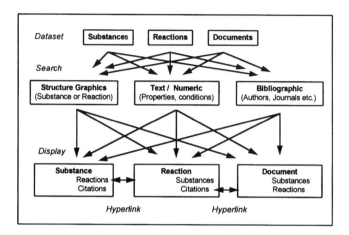

Figure 6.8. Relationships between the database objects Substance, Reaction, and Document.

- Details linked to a reaction include explicit graphical structures for starting materials as well as details of the reaction conditions, yield, etc., and each of the individual reports is linked to a literature reference to the document involved.
- Details linked to a document include details of the structures and reactions contained in that document, as well as the full bibliographic detail customarily used to index that document.

This symmetrical arrangement of three database objects is fully exploited in CrossFire: in principle, this means that operations on a data set of any of these types may be carried out in a search involving the context of any of the three, and the results are displayed (as a further data set) in the form of any of the three objects (Figure 6.8). The resulting display of results can then be used as a hyperlink from any embedded detail to a new window display containing the embedded object and all its further hyperlinks (and so on).

A few examples will illustrate this approach. As a first example, take the following case:

I seem to remember that Corey's synthesis of epibatidine involved a ring closure with nucleophilic attack of a nitrogen on a halide in the 4-position of a cyclohexane ring. Find me the reaction involved, since I want to find out how late in the synthesis path this ring closure took place. The reason for my search is that I want a synthesis path to an analog of epibatidine with the ring closure as late as possible in the synthesis, and Corey's work is a good starting point.

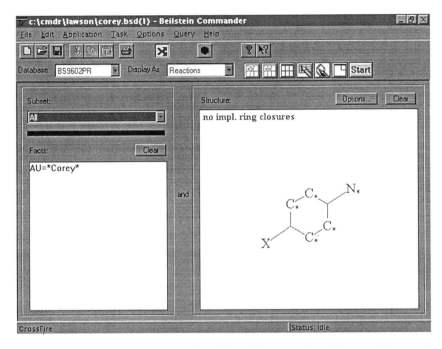

Figure 6.9. A mixed-context query involving Substance, Reaction, and Document data.

This is a fairly typical example of a real-life situation. The corresponding CrossFire query might look like Figure 6.9. In analogy to Figure 6.8, the following points should be noted:

- The starting data set to be searched is the whole database (BS9602PR). This could just as easily have been a subset (i.e., the hit set from a previous query in the same session, or a stored hit set from a previous session).
- The search is an intuitive combination of two components: a search for *documents* (AU(thor) = *Corey*) combined with a search for *compounds* (the graphical description). Both query components are equipped with powerful attributes to generalize the search (left and right truncation in the case of the author name, free valences and a generic term, X, in the structure). The ability to use these partial structure features in CrossFire with reasonable response times on huge databases gives CrossFire its power (this topic is discussed in more detail in Chapter 4).
- The results should be displayed as *reactions*.

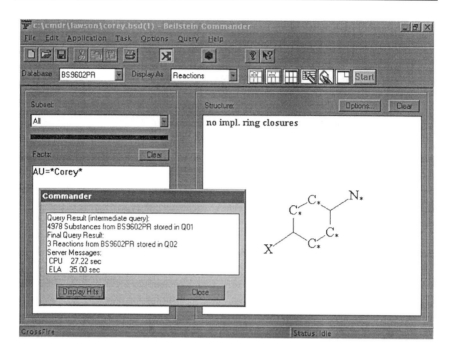

Figure 6.10. Response to the query of Figure 6.9.

This is therefore a concrete example where approximately 4 million *documents* were searched in the context of *substances* (there are approximately 7 million) and the results were displayed as *reactions* (approximately 5 million), in line with the nature of the topic as defined above. It is the aim of this system that this search be a completely natural request of the user, although in the background, as you may guess, the search mechanisms for such a mixed query are complex. The response of the system will look something like Figure 6.10, that is, although nearly 5000 substances were found in initial screening, only three reactions involving such compounds were attributable to Corey and his group. These have been stored in hit set Q02 as *reactions,* and we can look at them any time.

Examination of the hit set shows that the third hit is the step we were looking for (*see* Figure 6.11). While colors are not visible in the black and white reproduction shown in Figure 6.11, in the computer system the hyperlinks are color-coded: the black 6638620 denotes *substance* detail to the starting material, the blue 6637868 denotes *substance* detail to the product, and the green 5849673 denotes *document* detail to the Corey paper. In each case a mouse click opens an additional window of information, which itself includes further hyperlinks. In this way, it is very simple to trace a preparative path retrosynthetically, which we will pursue later in this chapter. In the meantime, we will assume that we hyperlink

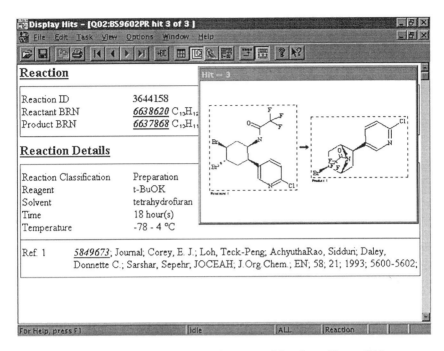

Figure 6.11. Browsing in the hit set resulting from Figure 6.10.

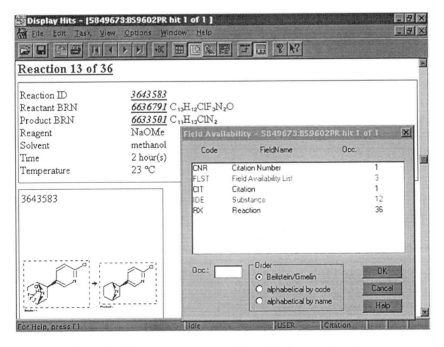

Figure 6.12. Result of opening a hyperlink from Figure 6.11.

to the Corey paper, set the User View to only reactions (there were 36 in the paper), and quickly scan through to find the final step in the epibatidine synthesis (Figure 6.12).

This result is enough to show us that the ring closure must be followed by selective removal of a bromine atom (in the presence of an essential chlorine atom) before removal of the trifluoroacetyl group. It looks tricky. So we decide to see whether any analogous ring closures are known, involving a later stage in the epibatidine synthesis. Corey's reaction shown in Figure 6.11 can be copied with a single click into the structure editor, where it can be modified quickly and resubmitted to the database as a query with the reaction sites marked and set to free valences (enabling leaving groups other than halogen to be found). Within a short time we have a hit (Figures 6.13 and 6.14), which shows that a methyl sulfonate in the correct position can bring about the ring closure in epibatidine in the last step (but in moderate yield, and we note that it was the racemate which was used).

Now we can trace the synthetic route backward via the Beilstein Registry Number (BRN), 6815573. One further click takes us to the nitro compound BRN 6819138, and we see immediately that this compound has (also) been prepared in a later publication by another group of researchers (Figure 6.15).

This is an example of the ability of CrossFire to trace synthetic pathways via single steps *over the artificial boundaries imposed by bibliographically based systems*. To the best of our knowledge, no other source of preparative and reaction information can emulate this investigative browsing: other systems that address multistep reactions are mostly based on the principle of cross-indexing the steps of a single paper at registration time, and are confined within the boundaries of the individual document.

This first example has shown us the methodology. In a short series, we have seen bibliographic, substance, and reaction searches combined with hyperlinking in an intuitive way to arrive at useful information in the recent literature.

The strength of the method lies in the speed and ease with which users are guided to the publications that interest them. As mentioned in the previous section, all true secondary information always leads to a list of publications that should be looked at by the user in greater detail in the original form. The merits of any individual system (such as CrossFire) lie mainly in how quickly and relevantly they enable the user to arrive at this list. This is partly a function of how quickly, graphically, and intuitively the system responds to the user's guiding hand (which we have emphasized in the first example), but it is also a question of how *deeply* the database content was indexed in the first place. Here the Beilstein database has a distinct advantage, because each compound and reaction in any particular paper has been indexed to a wealth of detail contained in numerical, text, and keyword fields.

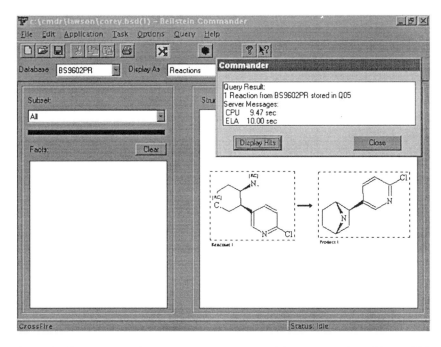

Figure 6.13. Redesign of the query context by editing a result from Figure 6.12.

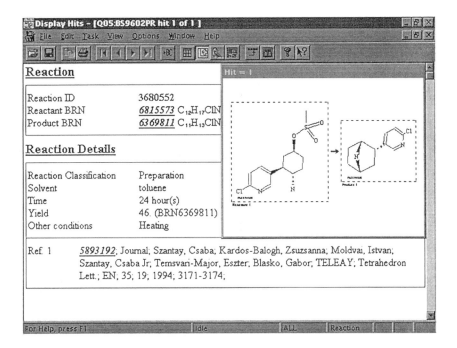

Figure 6.14. One result of the query of Figure 6.13.

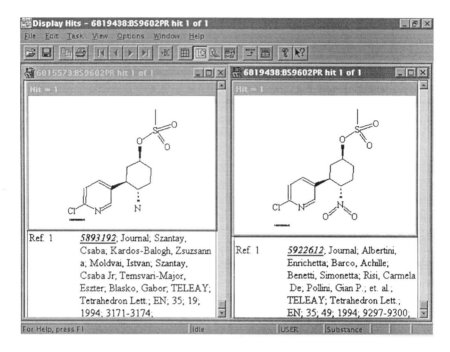

Figure 6.15. Retrosynthetic tracing finds another document.

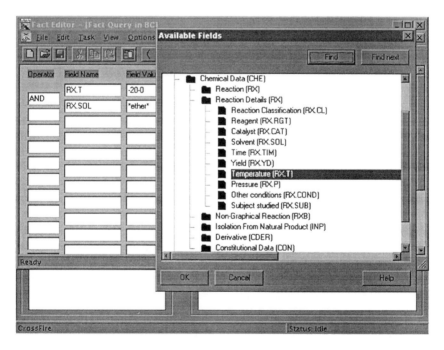

Figure 6.16. Browsing in the Beilstein data structure.

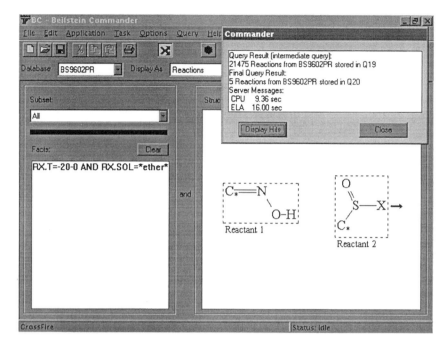

Figure 6.17. A reaction search using specific reaction conditions.

The second example we will discuss exemplifies the way CrossFire tackles topic information at a deeper level than merely structural graphics (although this is obviously a key feature). The case is the following:

> In a reacting medium of ether as solvent at low temperatures, I need to know what will happen if an existing oxime group is exposed to a sulfinyl halide in the reaction mixture. A sulfinyl oxime could possibly undergo a Beckmann rearrangement readily, and this would destroy my synthetic plan. Is anything known about this?

Here the question concerns experimental details of a nongraphical nature (reaction conditions). The user is assisted in finding the proper search term (out of about 400 such terms) by the use of a menu browser (Figure 6.16).

Note that the query is general (Figure 6.17): the graphics define all known oximes reacting with all known sulfinyl halides, but the reaction conditions are set to a range of –20 to 0 °C, and the solvent is any ether. The display of the five hits shows nothing unexpected from a chemical point of view (Figure 6.18). The five reactions are all from the same paper. We could have specified the document to be the display form at query time. Very often in reaction searching, we find tabular displays of many reactions in a single publication, so this is a useful option. However, we will now repeat this search in any case, and relax the condition about the

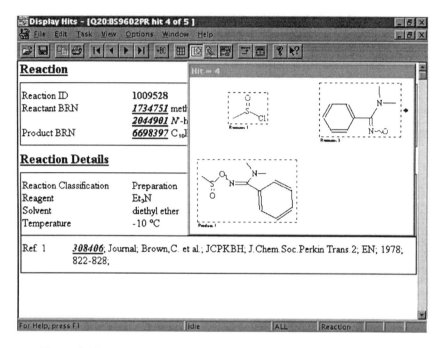

Figure 6.18. Display of a typical result from the query of Figure 6.17.

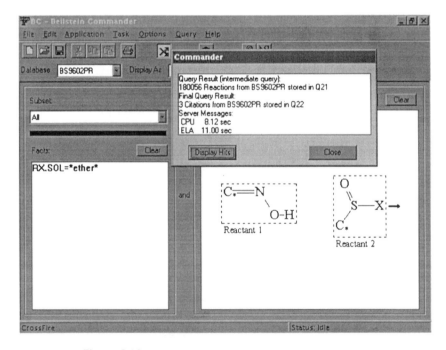

Figure 6.19. Modification of the query of Figure 6.17.

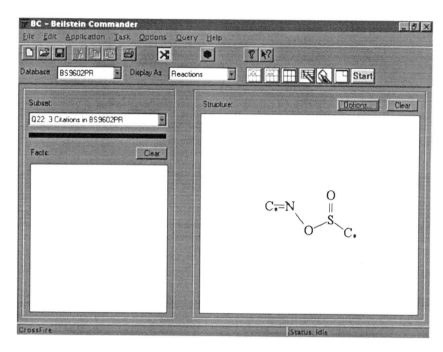

Figure 6.20. Searching the hit set produced in Figure 6.19 in a modified context.

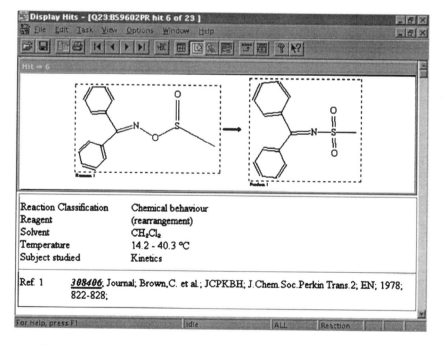

Figure 6.21. Display of a typical result from the search of Figure 6.20.

temperature, taking the opportunity to view the results in the form of documents. The results (*see* Figure 6.19) are interesting (three publications, of which the above is of course one). Now we can refine the query without having to go through all the individual data of these three key publications: the question is, what happens to these sulfinylated oximes as the temperature is raised or the conditions are changed in some other way (Figure 6.20).

Note that (1) we have chosen to search only on the content of the previous search (subset = Q22, the three citations), (2) we are searching for a particular set of compounds (the sulfinylated oximes), and (3) we want to see the reactions reported. This is a complex mixed-context query similar to the Corey example, but it is a quite natural demand on the side of the user, who need not be concerned about the mechanics of the system. The answer set (Figure 6.21) is interesting because it displays an unusual reactivity. The O-sulfinates rearrange at modest temperatures to yield the imine sulfonates (actually via a homolysis of the N–O bond), and there is consequently no Beckmann rearrangement involved. If we now repeat the search method without specifying any solvent or any temperature, the hit set widens to 11 publications in which the established method was applied by various researchers for various reasons, without the isolation of the sulfinylated oxime intermediate. This example shows how the indexing in CrossFire homes in on the essence of the topic—here, the exact conditions under which an oxime and sulfinyl halide will undergo this interesting conversion. In the same manner, the entire set of Beilstein property data elements can be used in any combination to formulate a query. This degree of detail cannot be dealt with adequately in this chapter; interested readers are assured that the functionality is entirely analogous, and they are referred to the CrossFire Users Manual.

The last example chosen here concerns the chemically simple topic of Friedel–Crafts alkylation of naphthalenes. The question is:

In the $AlCl_3$-catalyzed alkylation of naphthalenes, am I going to get simultaneous Wagner–Meerwein rearrangement of the alkyl halide that I use?

This is an interesting example because it illustrates the use of logical intersection of hit sets, a process which is necessary to separate the "sheep from the goats" when an exact definition of what you are looking for is hard to achieve in one query. This happens frequently, and CrossFire encourages the intersection of numerous hit sets in a completely uninhibited manner: as mentioned earlier, there are no system limits in CrossFire, so intersection of hit sets of even several million objects is possible. For example, (this NOT that) AND (one OR another) frequently become quite specific, even if one of these sets is very large. Newcomers to CrossFire are sometimes nonplused by the sudden lifting of the restrictions they

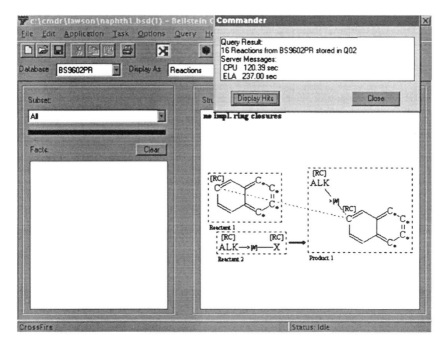

Figure 6.22. A reaction query for the alkylation of certain naphthalenes.

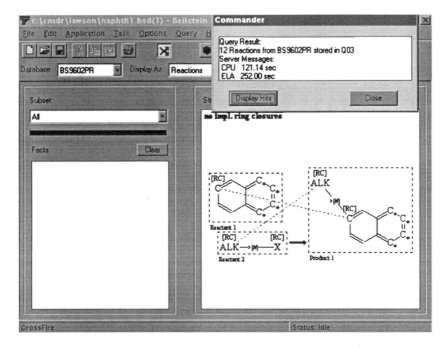

Figure 6.23. Modification of the query of Figure 6.22 to include the mapping of alkyl groups.

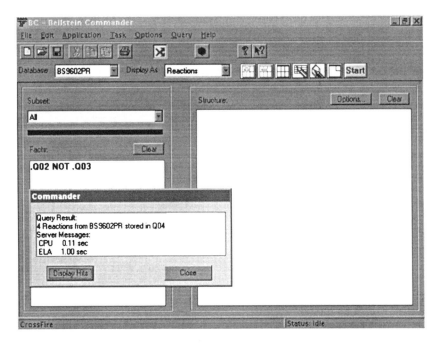

Figure 6.24. Hit set intersection for isolating the rearrangement during alkylation.

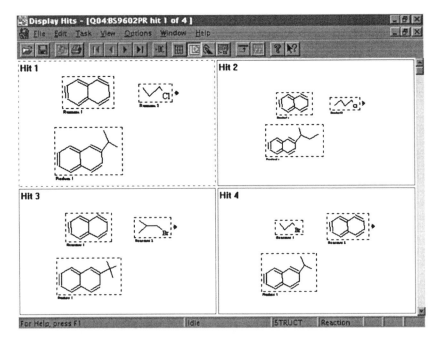

Figure 6.25. Overview of the reactions resulting from Figure 6.24.

have come to know in an online environment. In this example we are going to keep the numbers down for the purpose of didactic clarity.

As shown in Figure 6.22, the query is fully equipped with reaction center mapping, bond fate mapping, and one atom-to-atom mapping (between the beta positions of the starting material and product). Also note that this query took considerably longer (2 min of central processing unit (CPU) time, 4 min of elapsed time). It is important for the reader–user to realize that CrossFire can search for minutes in certain cases; the examples in this chapter were chosen to show a fair representation of the system.

The key point here is the use of the generic ALK (= alkyl), which in CrossFire is defined as "chain hydrocarbon" and can therefore be branched or straight. In other words, in the query as it stands, "alkyl" in the starting material need not be identical to "alkyl" in the product (it could have rearranged). In any event, the total number of reactions is 16, and they were stored in Q02. The next step is to repeat the search with the simple addition of an atom-to-atom mapping bond drawn from ALK to ALK; this ensures that "ALK" is interpreted identically on both sides of the equation, whether it is straight-chain or branched (Figure 6.23).

The search is no quicker, but the content of the hit set is smaller. There are 12 reactions, and they are stored in this case in Q03. Clearly, the difference between these sets is those four reactions in which a Wagner–Meerwein rearrangement took place under the conditions of the Lewis acid catalyst. And these reactions can be isolated simply by the last query involving the simple intersection Q02 NOT Q03 (Figure 6.24).

The four hits so obtained are all of the type expected (Figure 6.25). Obviously, with only four hits to check, this exercise could have been done manually, without intersecting the hit sets. But the principle is valid whether one is dealing with large sets or small: response times for intersection of the lists themselves are practically independent of the size of the lists in practice from the point of view of the user.

Future Directions

These examples of CrossFire at work have illustrated the basic concepts of the present system. Future development will concentrate on achieving additional indexing depth in areas where Beilstein has not previously ventured, in particular with regard to text searching and the inclusion of new data, such as ecologically and toxicologically important information.

References

1. Lawson, A. J., and Swienty–Busch, J. "A Comparison of On-line and In-house Performance," *Proceedings of the 17th International Online Conference, London, 1993;* Learned Information: Oxford, U.K., 1993; p 378.
2. The AltaVista search engine may be found at the URL: http://www.altavista.com.

Chapter 7

Using the Beilstein Reaction Database in an Academic Environment

Engelbert Zass

We have compared Beilstein CrossFireplusReactions with other organic reaction databases and systems, both in-house and public, that we have been using or plan to use in the near future at the Eidgenössische Technische Hochschule (ETH) Zürich. CrossFireplusReactions will be, along with MDL's ISIS, our major source of reaction information, supplanting both the tedious reaction searching done so far in the public Beilstein Online database, and the use of CASREACT and CA as premier "large" sources of information on organic reactions. The favorable experience accumulated in one year of using CrossFire, Beilstein's in-house database system for compounds and their properties, was an important factor in deciding to license its extension to organic reactions.

Introduction to Reaction Databases

The first public reaction database, CRDS (Chemical Reaction Documentation Service[1]) has been produced by Derwent since 1975 and made available via the host, Orbit, but it lacks the graphics that are essential for reaction databases even more than for other chemical sources. For query input, one has to use a tedious coding system for reaction partners and reaction types. Output for the time range 1946–1974 is in the form of references to the printed Theilheimer handbook only, to be looked up for the primary literature reference, and from 1975 onward, references are to both the printed *Journal of Synthetic Methods* (containing reaction diagrams) and the primary literature.[2] Fully graphical in-house systems such as REACCS (Reaction Access System, 1982),[3–5] SYNLIB (Synthesis Library, 1981),[4–6] and ORAC (Organic Reactions Accessed by Computer, 1983)[4,5,7] were developed not much later than CRDS, but they offered only a relatively small number of reactions in the beginning.[8] Public reaction databases with structure input and output appeared only in the second half of the 1980s and in the early 1990s: STN CASREACT[9] in 1988, STN ChemInform RX[10,11] in 1992, and STN ChemReact[11,12] in 1993.

© 1998 American Chemical Society

The utility of such reaction databases must be judged by both their information content and the means of accessing that information. Important criteria for the former are

- coverage, both by time and by primary sources excerpted,
- selection criteria for reactions,
- kinds of reaction data included (in particular, yields and reaction conditions),
- data types (e.g., numeric values such as yields or publication year as range-searchable numbers or as text) and their standardization,
- inclusion of multistep reaction sequences,
- reaction classifications, and
- size of the database, annual growth, and update frequency.

Criteria for means of access are

- whether substructure search capabilities are limited to reactants and products or also include reagents and solvents (and whether these are searchable only by name or also by internal and Chemical Abstracts Service (CAS) Registry Number),
- the quality of structure-mapping algorithms and reaction center recognition,
- the presence of similarity searching,
- the presence of facilities for searching subsets of the database,
- overflow conditions for very general queries,
- availability of role-specific "extraction" of information, clustering of results, import formats (structures, full reaction queries), and export formats (structures, data, tabulation of results), and
- integration and synergies with other databases.

The most obvious of these criteria are certainly size and time coverage of the database. Figure 7.1 illustrates, in a semiquantitative way, the size and growth of databases covering general organic reactions.[13] Two clusters are readily discernible. One includes relatively small and slow-growing databases, which might be called "selective" (i.e., selected reaction or reaction type) databases, which are characterized by a critical intellectual selection of the primary literature; this selection incorporates only the functional group transformations, formation of ring systems, etc., that are considered to be important, new, or representative examples. The other cluster includes the large and/or fast-growing databases such as Chem-Inform RX,[10] CASREACT,[9] and CrossFire*plus*Reactions; these are the "comprehensive" (or individual reaction) databases. Both clusters in Figure 7.1 contain databases, such as CLF[14] or *ChemReact*,[12] that are no longer updated but still play an important role within their respective clusters because of the time range they cover (*see* Figure 7.2).

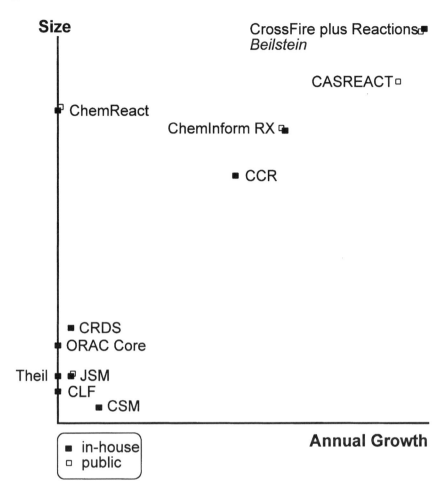

Figure 7.1. Clustering of reaction databases by number of reactions: size and annual growth (*see* reference 13). Examples (February 1997): CLF, 36,500 reactions, no longer updated; CRDS, ca. 82,000 (March 1997) + ca. 3000 annually; ChemInform RX, ca. 300,000 + ca. 50,000; CASREACT, ca. 1.3 million single-step reactions + ca. 40,000; CrossFire*plus*Reactions, ca. 5 million reactions + ca. 190,000.

This clustering also reflects the different applications of these reaction databases. Many reaction queries can be considered to fall in one of two groups: an organic chemist planning to perform a reaction in the laboratory will not need all the examples of the envisaged reaction type (and be flooded by that mass of information), but only representative examples that he or she can apply by analogy to the problem. For that kind of information, the chemist would turn to "selective" databases such as those in REACCS or its successor client–server systems ISIS/Host and

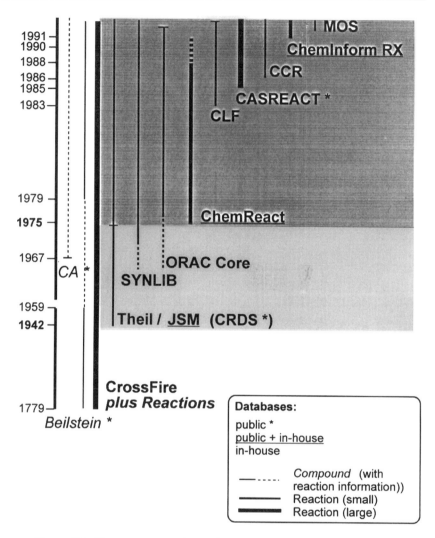

Figure 7.2. Time coverage of reaction databases (*see* reference 13).

ISIS/Desktop[15], or, if in-house systems were not accessible, to the public databases CRDS or DJSM[2]. If the attempted reaction and therefore the analogy fails, or if the chemist suspects from the beginning that standard reaction conditions might not work because of steric hindrance or other factors, or if the reaction is very unusual, then, in contrast, the chemist needs a large, comprehensive database. Only there can one expect to retrieve reaction examples closely related to the particular problem. This simplistic approach to the complex topic of reaction searching proved useful to us in evaluating reaction databases and making procurement decisions.

Obviously, an in-house system characterized by ease of use and a fixed searching cost, such as CrossFire*plus*Reactions, which covers both compounds and reactions, plays a central role in organic synthesis. But although we consider it an indispensable information source on organic reactions, we do not at present regard it as sufficient for all such queries; for the reasons given above and as illustrated later in this chapter, intellectually selected reaction type databases are still deemed necessary. We can only speculate at present as to (1) whether algorithmic extraction of reaction types, as pioneered by InfoChem in its database family of Chem-React and other subsets,[12] a feature later integrated by MDL into ISIS as a clustering tool,[15] would also work in as large a database as CrossFire*plus*Reactions, and (2) whether such a feature would make it a substitute for intellectually selected databases such as CRDS,[1,2] DJSM,[2] THEIL,[2,14] JSM,[2] or CSM[14].

Figure 7.2 shows that, because of the information from the Theilheimer handbook, reaction type databases extend back in time to the middle of our century (light shading), whereas the other "comprehensive" databases only cover the years from 1975 onward (dark shading). Obviously, with regard to time, CrossFire*plus*Reactions brought about a real breakthrough in coverage. Until its advent, the synthetic organic chemist had to access less-comprehensive reaction databases (with regard to both time coverage and number of reactions), or take recourse in compound databases such as CA. Both CA and Beilstein Online databases are not really amenable to full-fledged reaction searching by end users, because of the complexity and cost of the search process.

The Beilstein Information System

Beilstein has grown into an impressive information system, encompassing, besides the traditional printed handbook,[16] a database in several implementations and subsets (public online, in-house, CD-ROM), and ancillary software products, details of which can be found elsewhere in this book as well as in the published literature[17]. Figure 7.3 illustrates how the different components and media of this system can be used for searching numerous organic compounds and their properties, the traditional task for which Beilstein is well known.

Reaction information, as one of the "properties" of organic compounds, has always been included in the *Beilstein Handbook* and Beilstein database, although it could not be searched for directly in the printed *Beilstein Handbook*. Reaction information appears there predominantly under Preparation, indexed only at the product of a reaction, and only to a much smaller extent (*see* Figure 7.4) as Chemical Behavior, indexed at the reactant. In the Beilstein Online database, access to this information was improved. Reaction information was structured as shown in Figure 7.4 to make it searchable, and it was amended in a key aspect: whenever

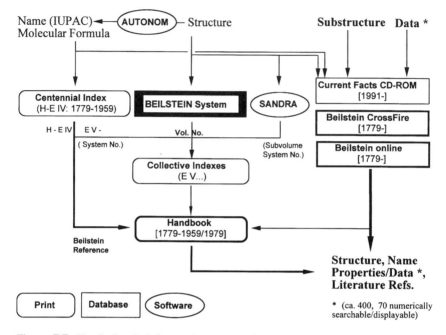

Figure 7.3. The Beilstein information system for searching organic compounds.

possible, names of reaction partners (starting materials in "product" records, products in "reactant" records) were automatically translated into the corresponding BRNs (Beilstein Registry Numbers) of the compounds that were added to the record, thereby *linking*, for the first time, compounds in their different roles in a reaction. However, this assignment of BRNs as a prerequisite for precise reaction searching was not always possible because some records contained, for reaction participants, only partial names, ambiguous trivial names, or molecular formulas.

In the predecessor to this book,[19] Damon Ridley described reaction data and its utilization in the Beilstein Online database in great detail. Ridley concentrated on reaction searches via substructure searches of product or reactant (and sometimes also on name or molecular formula searches, which were less expensive at that time), combined with names or name fragments of the other reaction partner, that is, product with "reactant names"/PRE.SM, or reactant with "product names"/REA.PRO.

As can be deduced from Figure 7.4, there is an alternative way, more precise and complete than such name searches, namely, using the BRNs assigned (as described earlier in this section) to link the result sets from (sub)structure searches for reactants and products in the database. Although we prefer this method of using BRNs instead of names, it suffers

Starting Material ⟶ Product

REA (8.5 %)	PRE (80 %)
Reaction product	**Preparation educt**
name [BRN]	name [BRN]
Reaction reagent	**Preparation reagent**
name	name
Reaction partner	**Preparation by-product**
name [BRN]	name [BRN]
Reaction detail	**Preparation detail**
text, name	text, name
Reaction subject	**Preparation yield**
text	%

Figure 7.4. Reaction information (data fields and type of information) in the compound records of the Beilstein database (*see* reference 18).

of course from the incomplete assignment of BRNs for reaction partners mentioned above. Therefore, to achieve the most comprehensive results possible, both approaches have to be utilized: a reactant names (fragments) search in the educt field PRE.SM, and BRNs from a reactant (sub)structure search requalified (using STN SmartSelect) as PRE.SM, have to be combined (using the Boolean AND) with a product (sub)structure search, thus linking two sets of "half-reactions", namely, all products prepared from the desired starting material(s) and all preparations of the desired product(s).[18] This utilizes the Preparation information in the Beilstein Online database only; to also use the Chemical Behavior data (REA, which is much less common, as shown in Figure 7.4, but mostly complementary to Preparation), additionally and conversely, the results of a reactant (sub)structure search must be combined with product names (fragments) in the field REA.RP and with BRNs from the product (sub)structure search requalified as REA.RP (*see* Figure 7.4).

Although this procedure is feasible for reactions of individual compounds (exact structures), or small groups of compounds, it is tedious in most cases involving substructures of a more general type. Until Beilstein abolished search charges in July 1995, extracting and requalifying a large number of BRNs were expensive. Our record-breaking case (but by no means the only one prohibitively expensive before this change in the price structure of Beilstein Online) occurred during a course in July 1994 when, fortunately, STN Karlsruhe had provided us with a free password. In order to retrieve direct hydroxylations of pyrroles to 4-hydroxypyrroles, a student (1) did a substructure search for product and reactant (at a total cost of about $110 at the full rate), (2) used SmartSelect for the reactant BRNs (6373!), and (3) then searched the extracted BRNs as PRE.SM (incurring a total of DEM 31,850 in search charges). Combined with the records from the product substructure search (436 compounds), the entire search gave four "reactions", that is, records of 4-hydroxypyrroles with references about preparation from a 4-unsubstituted pyrrole as starting material.

In the now obsolete version 2 of the CrossFire database, reaction information was placed into two data fields: Preparation and Reaction (for chemical behavior), as in the public database version of Beilstein (*see* Figure 7.4). With version 3 of CrossFire and version 1 of CrossFire*plus*Reactions, all such information is placed in the Reaction field if structures (graphics) are in the database for all starting materials and products. If this is not the case, the information is put into the Nongraphical Reactions field; this latter reaction information is not directly (i.e., via structures) searchable in CrossFire*plus*Reactions (for an indirect approach, *see* the next section). To compensate for the loss of precision produced by putting all these data in one field, a new field, Reaction Classification, was introduced, differentiating between Preparation and Chemical Behavior as with the former separate data fields. The Reaction field in CrossFire now contains reactions with the title compound as product as well as those with the title compound as starting material.[20]

CrossFire*plus*Reactions carried the transformation of what was originally an information source for compounds one step further: the BRNs assigned made the structures of reactants and products accessible and amenable to conversion into a "real" reaction database with (1) explicit, structurally searchable reactions, and (2) the added indispensable features of atom mapping and reaction center recognition for search and display. CrossFire*plus*Reactions is in the strict sense neither a new retrieval system—it is just an extension of CrossFire's substructure searching facilities with roles, mapping, and reaction centers—nor a new reaction database.[21] The reaction information, albeit only in implicit form, has been around, thanks to the traditionally thorough literature excerption, even in the printed *Beilstein Handbook*.

There are consequences of this "heritage" and the transformation that somewhat diminish the value of CrossFire*plus*Reactions with respect to reaction databases conceived as such from the beginning. Incomplete reaction information led to Nongraphical Reactions in the database, which are not structure-searchable because the automatic name conversion described could not be completed for all reactants and products. These reactions contain, nevertheless, useful information on the preparation of compounds or their chemical behavior, and can be located easily via hyperlinks (*see* the next section). We found that users are somewhat bewildered by these two types of reactions, so this point is important in training chemists to use CrossFire. Reagents and solvents are at present not searchable via (sub)structures, but only via names and formulas. Lack of standardization is a major problem here. For example, one finds Ti(O-iPr)$_4$, (i-PrO)$_4$Ti, titanium tetraisopropoxide, and some German besides the predominately English names. This is a significant disadvantage in such a large reaction database, where that type of information would be useful for narrowing down large hit lists (*see* the next section). On the other hand, CrossFire*plus*Reactions is not only the largest reaction database at this time, it exists also in the context of a large compound database with an unsurpassed wealth of physical data and properties.

Reaction Searching in CrossFire*plus*Reactions and Other Databases

The following examples[22] of reaction searches from the Eidgenössische Technische Hochschule (ETH) Department of Chemistry illustrate the use of CrossFire*plus*Reactions in comparison to both other in-house systems and large public databases.

The query shown in Figure 7.5 retrieved eight reactions in version BS9503 of CrossFire*plus*Reactions[23]; one of them is shown in Figure 7.6. The acetylation reagents used are given for all but one of these, which are classified as "reactant" as they contribute carbon atoms to the product, while as "reagents", pyridine (three reactions) and DMAP are shown; the last example is the only one that also specifies a solvent (CH_2Cl_2).

For comparison, a compound-based reaction search was attempted in STN Beilstein Online (last updated 3/10/96). The BRNs of 569 compounds retrieved with the reactant substructure were "extracted" and re-searched as preparation starting materials (STN SmartSelect; *see* Figure 7.4 and the discussion above) to give a total of 390 compounds prepared from any of those. The combination of these records with the 71 records from a product substructure search gave eight compound records that turned out to be identical to the reaction products and references retrieved in CrossFire*plus*Reactions (Figure 7.7). It should be noted that substituting Beilstein Online for CrossFire*plus*Reactions will work for such relatively well-defined

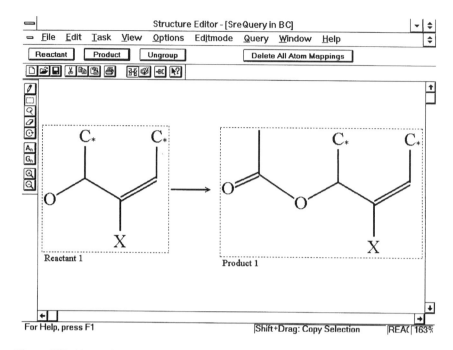

Figure 7.5. Example 1: Reaction query in Beilstein CrossFire*plus*Reactions (screen dump from Windows PC, Beilstein Commander PC Version 1.0). (Reproduced with permission from Beilstein Information Systems GmbH.)

queries. General queries like simple functional group transformations, or those needing atom mapping or specification of reaction centers, are not feasible using Beilstein Online. There is also the obvious increase in complexity when going from the in-house system to the public online database.

A corresponding query did retrieve one reaction in our REACCS installation,[14] while public databases at STN gave 5 reactions/3 references (CASREACT[9]) and 3 reactions/1 reference (ChemInform RX[10,24]). One reaction/reference from CASREACT was not considered relevant, as it involved acetylation of a cylcohexadiene ligand in an iron complex. The search in CASREACT was complicated by the fact that the reaction substructure query (equivalent to that shown in Figure 7.5) would not run to completion online but only as an overnight batch job. As in similar cases, we had to run this query not on the full database but on a subset (7215 references) created before by the Functional Group Term query SECONDARY ALC/FG.RCT (S) CARBOXYLATE/FG.PRO. These results point to a difference between the reaction databases examined here. While searches in REACCS databases give reactions with accompanying references (sometimes more than one), searches in CASREACT or STN ChemInform RX yield

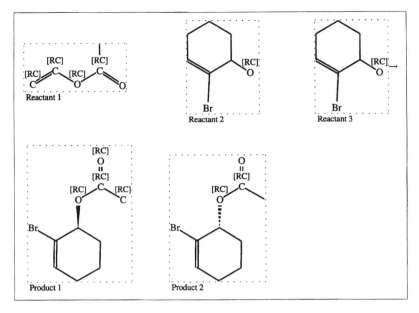

Reaction

Reaction ID	**1697103**
Reactant BRN	*1209327* acetoxyethene
	1926057 2-bromo-cyclohex-2-enol
Product BRN	*5477769* $C_8H_{11}BrO_2$
	5477770 $C_8H_{11}BrO_2$

Reaction Details

Reaction Classification	Preparation
Time	10 day(s)
Temperature	45 °C
Other conditions	Mucor miehei lipase

Note 1	Title compound not separated from byproducts

Ref. 1	*5654626*; Journal; Carrea, Giacomo; Danieli, Bruno; Palmisano, Giovanni; Riva, Sergio; Santagostino, Marco; TASYE3; Tetrahedron: Asymmetrie; EN; 3; 6; 1992; 775-784;

Figure 7.6. Example 1: Reaction retrieved in Beilstein CrossFire*plus*Reactions (printout on Hewlett-Packard LaserJet 4si). (Reproduced with permission from Beilstein Information Systems GmbH.)

literature references containing one or more of the reactions searched for. CrossFire*plus*Reactions can be switched from one of those "contexts" to the other: when queries are searched in their default context, structure queries retrieve compounds (with properties, including reactions, and the corresponding references), and reaction queries retrieve reactions (with reaction diagrams, reaction information, and references). When example 1

```
Rltd. Stereoisomers (RSI):    1948207; 5477769
Formula Weight (FW):          219.08
Lawson Number (LN):           5084; 1155

Ring System Data:

Number of Rings (CNR):        1
Ring Systems (CNRS):          1
Diff. Ring Systems (CNDRS):   1
Ring Heteros (CNRH):          0
Acyclic Heteros (CNAH):       3

Beilstein Ring Index | Ring System Formula | BRIX
(BRIX)               | (RF)                | Count
=====================+=====================+=======
 6.1.0-0.0-1.2       | C6                  | 1
```

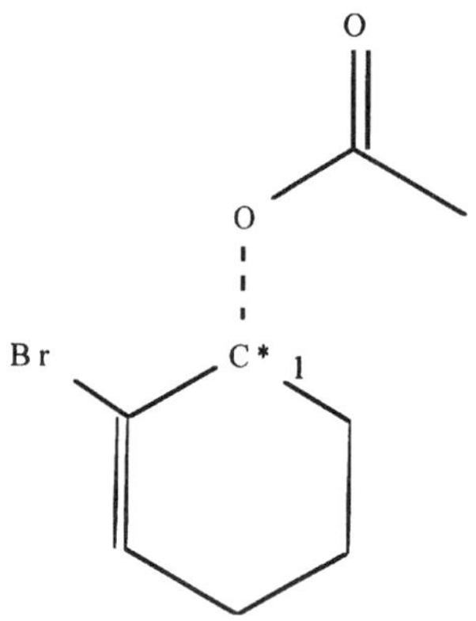

```
Atom/Bond Notes:
    1. CIP Descriptor: R

Preparation:
PRE
    Start:  BRN=1926057 2-bromo-cyclohex-2-enol, BRN=1209327
            acetoxyethene
    Time:   10 day(s)
    Temp:   45.0 Cel
    Detail: Mucor miehei lipase
    ByProd: BRN=5477769 C8H11BrO2
    Reference(s):
    1. Carrea, Giacomo; Danieli, Bruno; Palmisano, Giovanni; Riva,
       Sergio; Santagostino, Marco, Tetrahedron: Asymmetrie, 3
<1992> 6,
       775-784, LA: EN, CODEN: TASYE3
    Note(s):
    2. Title compound not separated from byproducts
```

Figure 7.7. Example 1: Product retrieved in STN Beilstein (search session captured in STN Express 3.21, Macintosh version; this is the same example as in Figure 7.6). (Reproduced with permission from Beilstein Information Systems GmbH.)

RX(2)   OF   3

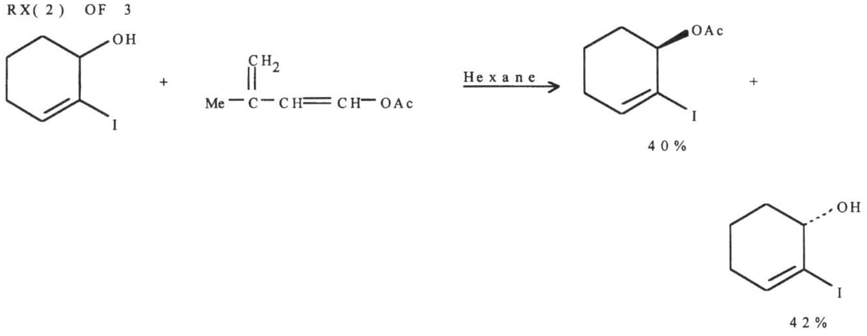

REF: Synlett, (10), 813-16; 1992
NOTE: Candida antarctica lipase B catalyzed

Figure 7.8. Example 1: Reaction retrieved in STN CASREACT (search session captured in STN Express 3.21, Macintosh version; this is the same example as in Figure 7.6).

was searched with the identical query, but in the Citations context, instead of eight reaction records (with citations), the corresponding six citations were retrieved, with all compounds and reactions abstracted from them on display, not only that searched for.

The overlap between systems and databases in example 1 was as follows. The reaction from REACCS (1981, InfoChem Reaction Database[12,14]) was not found in any of the other databases, whereas the one reaction from ChemInform[24] and a different one of the eight from CrossFire*plus*Reactions were also retrieved in CASREACT (Figure 7.8). Of the eight reactions in CrossFire, four originated from the literature before 1985 and could therefore not be retrieved in CASREACT. Although the two references shared by CrossFire and CASREACT and by CASREACT and ChemInform RX are different publications originating from different research groups, they describe the same method for the desired transformation, using a lipase. This illustrates both the problems of defining overlap between reaction databases precisely (on the basis of references, reaction examples, or reaction methods), and the usually small overlap between such reaction databases, due to different time coverage and selection policies.

Searches for selective transformations of one functional group in the presence of others are as easy in CrossFire*plus*Reactions as in other reaction databases. For the selective reduction of cyclohexanone esters to cyclohexanone alcohols (example 2, page 112), four reactions were found in CrossFire*plus*Reactions (two with one, the other two with two references each, for a total of three different references from 1992, 1991, and 1986). All of those were annotated as multistep reactions; such multistep reactions in CrossFire are searchable either as only the overall reaction, as in these cases, or as individual steps. Here, CrossFire*plus*Reactions is somewhat at a disadvantage to REACCS/ISIS, where both individual steps and the overall reaction are searchable. It is at a particular disadvantage to CASREACT, where not only are individual steps (provided the intermediates were isolated or at least characterized) and the overall reaction searchable, but also all partial recombinations of the steps are retrievable.

A search for the direct conversion of benzyl alcohols to *S*-benzyl isothioureas (i.e., not via halides prepared from the alcohols; example 3, above) gave nine reactions (four different references: 1956, 1958, and 1987) in CrossFire*plus*Reactions, but only one in CASREACT, which was obviously missed in CrossFire because it involved a one-pot, two-step reaction via the benzyl bromide not desired in this case. ChemInform RX[24] gave no answer at all to that query.

Unusual functional group transformations such as the last example, or a common functional group transformation (e.g., formation of an acetate) that is embedded in a "special environment" (e.g., allyl position, halogen substituent) as in example 1, usually need large reaction databases. In contrast to the searches shown so far, example 4 (page 113) is in its most general form a simple, common functional group transformation where large databases give too many answers to look at in detail. Smaller, "selective" databases, however, can be expected to give a manageable number of representative examples for the transformation. Table I shows this difference between "comprehensive" reaction databases such as CrossFire*plus*Reactions, and "selective", reaction type databases for several variations of a query that differ in the extent to which the substructures of the reaction partners are specified.

As can be seen in Table I, even the most restrictive case (with R = phenyl) gives more reactions in the large databases than a chemist would

$$\text{F}_3\text{C} \overset{\text{O}}{\underset{}{\parallel}} \text{R} \longrightarrow \text{F}_3\text{C} \overset{\text{OH}}{\underset{}{|}} \text{R}$$

Table I. Reactions Retrieved for Reduction of Trifluoromethyl Ketones

Reaction database	CrossFire	CASREACT	REACCS
Coverage since:	1779	1985	1946
No. of reactions present (million)	5	1.3	0.45 (14)
R = C (in a chain only)	161	118	
C (in a ring only)	134	100	15
Phenyl (subst.)	99	82	8

normally care to look at. To find general methods for such a transformation, a database system with smaller, intellectually selected reaction databases such as REACCS or ISIS would be preferable. All searches in Table I were done with substructures and their roles (reactant, product) in the reaction. Such simple queries often retrieve false hits that contain the required substructures but not the required transformation. In order to enhance precision, reaction databases must include information on the correspondence of atoms in reactants and products (atom-to-atom mapping), thereby identifying reaction centers (bonds and/or atoms) in a formal (not a mechanistic!) way. Mapping and reaction center recognition are usually done algorithmically during creation of the database. For queries, some systems allow only manual mapping and/or reaction center marking (CrossFire*plus*Reactions, CASREACT), whereas others also permit algorithmic mapping in queries by the same algorithm used in the database (REACCS, ISIS). Mapping and reaction center marking are obviously useful precision tools, but they have their pitfalls in practice: the query with "R = carbon in a chain" retrieved 161 reactions in CrossFire*plus*Reactions; this number was reduced by either mapping, reaction center marking of the C=O bond in the starting material, or a combination of both to 127 reactions each. The missing 34 reactions, however, were practically all considered to be relevant, but they were not retrieved with the refined queries because they lacked mapping information (and therefore reaction centers as well) in the database.

This problem is not particular to CrossFire*plus*Reactions, as example 5 (page 114) shows. For this epoxidation, we retrieved 1177 reactions in CrossFire*plus*Reactions when searching with roles, but only 818 reactions when also specifying mapping and reaction sites. The corresponding results for the same query were 1032/914 in CASREACT,[9] 180/163 in STN ChemInform RX,[10,24] and 55/49 in REACCS[14]. Four reactions retrieved in

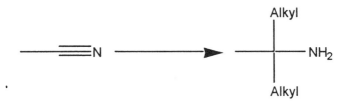

REACCS with "roles only" were false hits, and two were correct answers missed with the mapping. Spot checks for the reactions missed by searching with mapping and reaction centers in both CASREACT and Cross-Fire*plus*Reactions indicated that the majority of those missed seemed to be relevant. This prompts us to recommend mapping routinely in REACCS or ISIS, but only if really needed for precision in CrossFire*plus*-Reactions or CASREACT. From a user's point of view, this situation needs improvement.

Example 6, above, conversion of a nitrile into a tertiary alkylamine, gave, in CrossFire*plus*Reactions, 107 reactions with substructures and roles alone (query 6a), 58 with roles plus the C–N triple bond marked in the query as a reaction center (query 6b), and only 38 by the most precise query with the same reaction center and additional mapping of the carbon atom bound to nitrogen (query 6c). Examining the different results from queries 6a and 6b, we found that of the 49 reactions missed by query 6b, 20 had no reaction center information at all (because they had no mapping), but 5 of those were relevant; the remaining 29 with mapping available in the database (9 of those with the nitrile group as reaction center) were not relevant. The 20 reactions retrieved by query 6b but not by query 6c were all not relevant, as they had the amino group already present in the starting material, or generated it from an acetylated derivative, while the nitrile group reacted to an amide or was reduced to a primary amine. In sum, specifying both the reaction center and mapping in the search eliminated 64 irrelevant, but also five relevant reactions!

An important and unique aspect of CrossFire*plus*Reactions is the context in which it puts reaction searches. By virtue of the impressive compound and citation hyperlinks provided, the full range of property and literature information is available for products and reactants at a mouse click and is presented in a form that is user friendly. Likewise, an individual reaction searched for can be put into the context of all other reactions and compounds for every reference retrieved by activating a citation hyperlink. Vice versa, compounds and citations found in respective searches can be used as jumping-off points to reactions. Figure 7.9 is an attempt to capture the information structure underlying these hyperlinks. Some links are of a 1:1 type—for example, activating a product hyperlink in a reaction leads to a single record for that particular compound—multipronged arrows in Figure 7.9 symbolize 1:many links, as a citation hyperlink from either reaction or compound usually leads to several compounds and reactions contained in that publication. All hyperlinks are color-coded for easy navigation: reactions are red, products blue, starting materials black, and citations green. Thus, browsing and serendipity are supported as in no other database known to the author at present. Although hyperlinks provide immediate changes between the different views—or, as named by Beilstein, contexts: Substances, Reactions, and Citations—of the same database after any type of search in CrossFire*plus*Reactions, setting the context before the search displays results in the desired view immediately.

For example, one of the reactions retrieved in the search for example 4 (Table I) was the reduction of a trifluoroacetophenone to the corresponding (*S*)-phenylethanol with a chiral Grignard reagent. Clicking on the hyperlink representing the product brings the corresponding compound record onto the screen, showing (1) physical data such as melting and boiling points, refractive index, and optical rotation (all with numeric values, 12 for the last property), (2) references for circular dichroism (CD) and optical rotatory dispersion (ORD) spectra, and (3) 10 reactions and one Nongraphical Reaction, that is, a reaction without full structural information for both starting materials and products (as discussed in the preceding section). In this example, it was a preparation of the title compound with the starting material information missing from the database. For such "reactions", the original literature needs to be consulted to determine whether this is actually the desired reaction or a preparation from a different starting material. If one is only interested in preparations of a compound, one can activate the filter Preparation in the View menu[20]: in this example, only five of the original 10 reactions (and the Nongraphical Reaction[20]) remain on display, and the numbers in the Field Availability list are adjusted accordingly. When activating a citation hyperlink here, the user gets all compounds and reactions abstracted from the publication (including the one originally found in the reaction search), optionally with the appropriate reaction diagrams (Include Structures, which is in the View pull-down menu). Hyperlinks can also often be utilized to "roll back" synthesis sequences to a commercially available starting material, as catalog information from Aldrich, Fluka, and Merck is included in the XREF field of substance records in CrossFire.[25] Although linkages from reactions to additional information are also possible in other reaction databases discussed here, they are more tedious and give less-perceptive results. Costs for such operations in the public reaction and compound

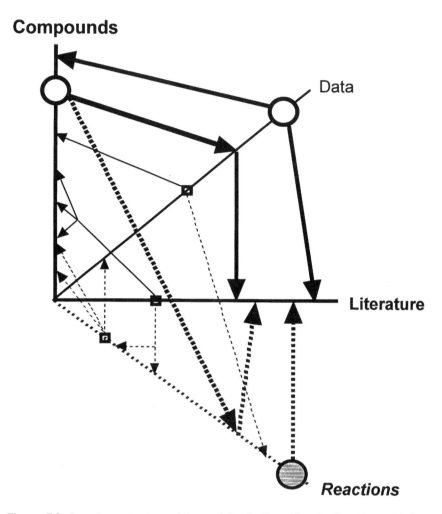

Figure 7.9. Search contexts and hyperlinks in CrossFire*plus*Reactions. Circles: query types; heavy arrows: type of information retrieved by a query (solid lines: CrossFire; dashed lines: CrossFire*plus*Reactions only); rectangles: starting points for hyperlinks; light arrows: type of information accessed by a hyperlink query (solid lines: CrossFire; dashed lines: CrossFire*plus*Reactions only).

databases at STN act as an additional deterrent.[26] No database available at present other than CrossFire*plus*Reactions gives us organic compounds in three different contexts—substances, reactions, and publications—with a mouse click.

While the examples shown so far cannot of course claim to be representative, they can be considered indicative of the fact that the reaction databases discussed here serve their purpose well, and some are obviously superior in ease of use and in results retrieved. The whole armory of databases may have to be used to achieve comprehensiveness, and, although the majority of reaction searches run successfully, as shown here, we will now come to two further examples that are definitely more problematic with most existing sources and facilities.

For reaction queries with common reactant and product substructures, searchable reagents, catalysts, and solvents are desirable if not essential. In order to find functional group transformations with unusual reagents and solvents, or special application of common ones, all reaction partners besides reactants and products must be fully searchable via (sub)structure. REACCS, ISIS, and STN CASREACT/Registry permit (sub)structure searches for all reagents and solvents, while CrossFire does not at present because these reaction partners are stored only by name or molecular formula, and not by structure. In example 7, conversions of benzyl ethers to the corresponding acetates with acetic anhydride and trimethylsilyl triflate as reagents were desired. A reaction search with a benzyl ether substructure and acetic anhydride as reactant (as it contributes carbon atoms to the product, it is formally a reactant, not a reagent) and acetates as a product substructure gave 3500 reactions. An attempt to narrow that down as desired with "RX.RGT="trimethylsilyltrifl*"", using the mask-driven Fact Editor of CrossFire with excellent help and database-lookup features for text entries, gave zero answers. Unfortunately, this text search for reagents and solvents is not as reliable as a structure search. In CASREACT, such a query can be searched for more precisely using substructures for starting materials and products, and CAS Registry Numbers retrieved in STN Registry for reagents, but the result in this example was also zero.

In addition to being searchable via structure, all reaction partners, not just starting materials and products, must also be "extractable" from reaction search results in a role-specific way. Reaction conditions such as temperature, pressure, pH, and yield should also be "extractable", with tools to tabulate all this extracted information for easy and fast perception. CrossFire unfortunately lacks such features, while REACCS Conversion Searches,[27] and, more tediously and without tabulation, STN SmartSelect[26] in CASREACT, permit such "extractions". Such features are already desirable in small databases, but they are essential for large "comprehensive" databases with their concomitant large hit lists for generalized reaction queries. Only such tools, for example, would make large numbers of reactions retrieved (e.g., as in example 4) useful at all, as chemists could just extract from the many examples the different reagents and reaction conditions used.

Although general "type of reaction" queries are normally searched for in the appropriate "selective" databases, searching large databases such as CrossFire*plus*Reactions or CASREACT becomes mandatory when only special instances (as defined by particular reagents, conditions, etc.)

of a structurally general reaction type with many common examples are needed. A search for nonstandard methods to deprotect *tert*-butoxycarbonyl (BOC) amino acids (example 8) retrieved 55 reactions in REACCS,[14] which were not examined in detail but from which 24 different reagents were extracted in a so-called Conversion Search and displayed as a table. A similar approach for the 1643 reactions retrieved in CrossFire was not possible, making these search results virtually useless. In CASREACT, 900 references with 11,535 reactions were retrieved by a reaction substructure search and narrowed down to single-step reactions (814 references). Eliminating those with the standard reagent trifluoroacetic acid left 517 references that were linked to the keyword "DEPROTECT?", to give only two references, one using trifluoroacetic anhydride as a reagent, and the other one a false hit (a multistep one-pot reaction).

We decided then to go back to the 517 references and to SmartSelect all reagents from the "hit reactions". These 170 reagents were displayed in the free format SCAN IN (*Index Name*) in the STN Registry file, and evaluated for potential usefulness; 36 reagents considered to be of interest were linked with the saved original 517 references in CASREACT to yield 85 references. Evaluating these by DISPLAY SCAN and eliminating HCl and acetic acid as reagents finally gave 50 references that were printed out (formats SCAN in CASREACT, and CBIB in CA)—without good tools, reaction searching such as this can be tedious.

Facilities for specifically extracting information from reactions retrieved are obviously useful for refining and further searching, and important tools for postprocessing retrieval result. Postprocessing should also include facilities to sort and cluster reactions by structure types, by reaction centers, or by reagents, solvents, and reaction conditions in order to enable the management of hit lists, particularly from large databases such as CrossFire*plus*Reactions or CASREACT. With its clustering, ISIS[15] is presently the most advanced system in this respect, whereas CrossFire*plus*Reactions lacks such features at present that would be useful and necessary, particularly for this database.

So far, all of our examples have involved substructures for reactants and products. In the next three examples, exact transformations are sought. As the probability of finding the desired reaction in one of the small, selective databases is by definition low, this type of reaction query must be addressed in large databases such as CrossFire*plus*Reactions and CASREACT.

A search for the Baeyer-Villiger-type oxidation of norbornanone (example 9) to the two possible products uses CrossFire's generic search features[28] in an exact compound search to avoid having to specify two separate queries for both cases (Figure 7.10). Six reactions with a total of 12 different references were found. Corresponding searches retrieved seven reactions/references in CASREACT, four reactions/references in ChemInform RX,[24] and none in REACCS[14]. Of the 15 references retrieved

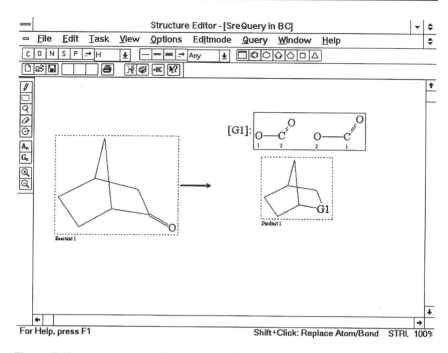

Figure 7.10. Query in CrossFire*plus*Reactions for example 9. (Reproduced with permission from Beilstein Information Systems GmbH.)

as a total from these three databases, only one was common to all of them; two were shared by CrossFire/CASREACT, and three each by Cross-Fire/ChemInform and CASREACT/ChemInform, respectively. One reference found exclusively in ChemInform is actually a preliminary publication (*J. Chem. Soc. Chem. Commun.* **1993**) for which the full paper (*J. Chem. Soc. Perkin. Trans. 1* **1994**) was found in both CrossFire and CASREACT. Two references can be accounted for being exclusive to CAS-REACT, one a 1995 Japanese patent,[29] the other from *J. Mol. Catal. A*, a journal covered only sparingly by Beilstein (24 references in the entire database, none later than 1980), and not at all by ChemInform RX. Although three of the relevant references found exclusively in CrossFire antedate the time covered by CASREACT, another three missed by the latter database cannot be accounted for by known coverage policies; among them is a 1994 publication in *J. Org. Chem.* that was also missed by ChemInform RX.

In example 10, a query for the oxidation of styrene to the corresponding diacetate (1,2-diacetoxy-1-phenylethane, stereochemistry not specified) retrieved the following numbers of references: 14 in CrossFire*plus*Reactions (reaction structure search); 7 in CASREACT (via the CAS Registry Number of styrene combined with those of the four products registered by CAS, plus the appropriate roles in the reaction: /RCT, /PRO[30]); and 2 in ChemInform RX[24]. All references from the other two databases were also retrieved in CrossFire, which exclusively gave four references from 1967, 1980, 1984, and 1987. Remarkably, three references included in CrossFire (*Tetrahedron* **1991**, *J. Chem. Soc. Perkin Trans. 1* **1985, 1983**) were missed in CASREACT but were retrieved with a similar query (CAS Registry Numbers + roles) in the STN CA file; the 1991 publication was also retrieved from ChemInform RX. As far as individual reactions in those references are concerned, there were 26 (CrossFire*plus*-Reactions), 11 (CASREACT), and 2 (ChemInform RX), respectively. Analysis of the synthetic methods used for this transformation revealed nine different methods in CrossFire*plus*Reactions, six in CASREACT, and one (electrochemical oxidation with Mn(III), also found in the other two databases) in ChemInform RX. Exclusive to CrossFire were oxidations with iodine, Co(III) chloride, or ruthenium chloride plus peroxides, and exclusive to CASREACT was diphenyl ditelluride as a reagent (while both CrossFire and CASREACT had two examples each using benzenetellurinic acid as an oxidation reagent). Lead tetraacetate was utilized in six examples in CrossFire*plus*Reactions (four Preparations, two classified as Chemical Behavior), and in two from CASREACT.

Searching for the synthesis of aniline by direct amination of either benzene or phenol (example 11) illustrates some problems with CrossFire*plus*Reactions: a query with exact structures and their roles retrieved 29 reactions using benzene as a starting material and 10 reactions using phenol as starting material; with mapping of the carbon atom bearing the amino group in the product, these numbers were reduced to 8 and 1, respectively. Among the eight reactions involving benzene, only two were relevant (with a total of eight references), and the other reactions were formally correct but contained additional reaction partners and were not showing the desired transformation. The one reaction retrieved starting from phenol was relevant (the other nine retrieved without mapping were not), including a patent (1930), three journal references (1934, 1980, 1991), and a reference each to Ullmann and Kirk–Othmer.

Although routine use of mapping in CrossFire*plus*Reactions is not recommended for the reasons given earlier in this section, and mapping (or reaction centers) are normally not necessary anyway when looking for "exact" reactions, in cases such as this with common compounds, it has to be considered as a precision tool. CASREACT gave seven reactions/references for the amination of benzene: three of the four from CrossFire that were published after 1985 (the one missed in CASREACT was classified in CrossFire as "Subject studied: Mechanism"), plus two patents[29] and two journal references not found in CrossFire (one of the latter was the only one retrieved from ChemInform RX). Six reactions/references were found in CASREACT for the amination of phenol; none of these four patents[29] and two journal references (*Bull. Chem. Soc. Jpn.* **1985**, *Dokl. Akad. Nauk. SSSR* **1988**) were retrieved in CrossFire, and no such reaction at all could be found in ChemInform RX. All journal references found in CAS-REACT but missed in CrossFire for both aminations were from journals that are covered by Beilstein in principle.

Although it is of course impossible to make general conclusions about database quality and coverage from the few examples discussed here,[31] they at least indicate that CrossFire gives good enough coverage for reactions to make it a first (and sometimes only) choice for many searches, excepting of course those queries such as examples 4 and 5 (maybe also 6 and 8), where selective databases (*see* Figure 7.1) are preferable. The question remains as to whether it is advisable to reduce the sometimes large numbers of hits in the "comprehensive" databases such as Cross-Fire*plus*Reactions by more specific, narrower queries. For us at the ETH, this would imply "reeducating" our users, who have been told so far to phrase reaction queries broadly in the "selective" databases that have been available to them at their workplaces via REACCS, SYNLIB, and ORAC since 1985. For searches that need to be as comprehensive as possible, additional sources must be included, obviously CASREACT,[9] in particular for patents after 1991 and less-common journals, but also ChemInform RX,[10] ChemReact,[12] and the selective in-house or public databases, as even in searches for exact transformations they might provide a "lucky hit".

Finally, it is worth noting that some unusual queries are still searched better in a general chemical database such as the CA file than in specialized reaction databases. A recent search for organic reactions in liquid nitrogen as a solvent retrieved nothing in CrossFire*plus*Reactions and REACCS,[14] and nothing useful in CASREACT: "7727-37-9/SOL" gave 2536 references, but these seemed to use nitrogen as a protective atmosphere according to some examples examined, and linking those to "LIQ?/NTE" (76 references) gave zero results. In the CA file, the query "(LIQ? (A) (N2 OR NITROGEN)) (S) (SOLVENT? OR SOLUB?)" turned up 72 references, some of which were examined (DISPLAY SCAN) and then refined by combining them with the CAS Registry Number for nitrogen to give 34 references; most of these were relevant.

Reaction Databases at the ETH Zürich Chemistry Department

Since late 1985, organic chemists at the ETH Zürich Chemistry Department have had access to reaction databases offered by the systems REACCS,[14] ORAC (until 1992), and SYNLIB. These databases were loaded on VAX computers running under the VMS operating system in the ETH Zürich Computer Center, and made available at the chemist's bench via the ETH LAN KOMETH (19,600 bps), later also via Ethernet. Access is almost exclusively via the Tektronix 4105 graphic terminal emulation provided by VersaTerm-PRO on Macintosh computers. Newly acquired database systems such as CrossFire, ISIS, and SpecInfo are now loaded on Unix servers under the direct supervision of the chemistry library, which is really already the Chemistry Information Center, with electronic sources, both in-house and public, continually gaining in importance. This change in responsibility was caused by decentralization of computing resources in our institution, induced by the availability of relatively cheap and powerful servers.

Client software must be available for both Macintosh and Windows PCs, as we have a mixed population of those machines in our department, with one type usually dominating in an individual research group or institute. Access to such client–server systems is via Ethernet and the TCP-IP protocol, which is supported not only in the entire ETH Zürich network, but also worldwide via the Internet. This protocol therefore also enables sharing access to such databases among several academic institutions in consortia that will not only reduce the overhead for operating servers—a particular problem in small universities—but also lead to economically more favorable license conditions than do individual licenses.[32]

For the future of reaction information at our department, we decided at the end of 1995 to replace REACCS[14] by MDL's ISIS/Host, and to extend our successful CrossFire[33] installation to CrossFire*plus*Reactions, for the following reasons. First, REACCS will not be supported indefinitely by MDL. Second, ISIS is standard in many chemical companies, and ISIS offers features and databases not available on the older system. In addition, ISIS will run on the same Unix server (IBM RS/6000) as CrossFire, obviating the need to take care of both Unix and VMS installations with our limited staff. Because of the significantly different size and origin of reaction databases in ISIS and in CrossFire*plus*Reactions, and of the different search facilities illustrated above, we think that we need both a "comprehensive" database such as CrossFire*plus*Reactions, and the "selective" reaction databases provided by REACCS and ISIS. As organic reaction searches are everyday questions in our department, licensing in-house systems makes sense economically, particularly at the rates offered for academic institutions. It serves to remember here that of the public reaction databases, only ChemInform RX[10] is offered in the STN Karlsruhe Academic Program with significant price reductions, while CASREACT[9] and DJSM[2] have to be paid for in full.

The decision to license in-house systems was and still is also dictated by their superior user interfaces compared to corresponding public reaction databases. With these in-house systems available and routinely used, we will turn to public databases only as a "last resort". This will almost invariably be done by an information specialist consulted by the end user.

He or she will ascertain that the desired information cannot be retrieved otherwise—one should not forget that printed reaction compendia such as Houben–Weyl, Patai, and many others have by no means outlived their usefulness in the electronic age!

We expect our chemists to use CrossFire*plus*Reactions as a routine tool for searching organic reactions by input of reaction substructures. We also expect that after compound (sub)structure searches, the chemists will use the features in CrossFire*plus*Reactions to view reactions of the compounds retrieved.

Although CrossFire[33] is one of the most user-friendly systems we have seen so far, our experience with observing users in the first year[34] made it clear that a basic introduction to such a database is necessary if we want our users to profit from CrossFire as much as possible. As with all user-friendly information systems, the emphasis is less on handling the system than on illustrating the facilities and putting the system in the context of other sources. Training and support will emphasize the following important points. Users will be reminded to utilize ISIS also (or even preferentially) in cases where they get more results than they are willing to look at, as in more-common functional group transformations or when reagents and solvents are involved. While training for REACCS (and now ISIS), we had to correct the observed tendency of users to overspecify their queries. We will have to do the opposite for CrossFire, illustrating the differences between "comprehensive" and "selective" reaction databases. With regard to mapping and reaction centers, with which problems are seen not only in CrossFire, users are advised to use them only when they are really necessary to enhance precision or to narrow down large lists of hits.

We offer regular (once per month) one-hour introductory courses for CrossFire. Once both CrossFire*plus*Reactions and ISIS reaction databases are installed, there will be an additional one-hour introductory course (every fourth week during term time) on chemical reactions. In addition, there will be half-day in-depth courses with demonstrations and hands-on practice for both CrossFire*plus*Reactions and ISIS, like the courses we have so far been offering for REACCS two to five times per year. Courses will be offered more often in the beginning, depending on demand.

As with all in-house databases and other important chemical information sources, we provide individual assistance and additional courses tailored for research groups on an as-needed basis. The entire Beilstein Information System is covered in the one-year (two terms), one-hour chemical information course for graduate and advanced undergraduate students taught since 1984.[35] Important information on CrossFire, both technical and about context and use, is conveyed to users via our World Wide Web (WWW) pages.[36]

Conclusions

Our experiences with CrossFire[33] are such that we consider it an absolutely indispensable source of information about organic compounds and their properties. Even in this compound context alone, CrossFire*plus*Reactions could be considered a useful extension to easily view reaction information for compounds retrieved. But CrossFire*plus*Reactions becomes indispensable, too, for those organic reaction queries that need large databases, as there is no viable alternative at present in terms of time coverage and number of reactions (*see* Figures 7.1 and 7.2). With regard to the last criterion, only CASREACT can be considered competitive, but it is too limited in time to be a real alternative even if it were available in-house. Nor is CASREACT, even when accessed with the help of STN Express, considered user-friendly enough to be qualified as an end-user system.[37] On the other hand, the availability of CrossFire in its present form does not make systems such as REACCS or ISIS obsolete, because there is a definite need for the kind of intellectually selected smaller and specialized[2,12–14] databases on those systems.

Appendix
Operational Experience with Beilstein CrossFire

While this chapter was being written, we were upgrading disk storage on our server prior to installing CrossFire*plus*Reactions. After releasing CrossFire*plus*Reactions in October 1996, we found that the technical and organizational issues concerning installation, maintenance, and user support for this database were very similar to those experienced with the compound database system CrossFire.[33,34]

CrossFire has been running on an IBM RS/6000 7012 POWERstation/Server Model 380 (64 MB of RAM, two SCSI-2 controllers, SCSI-2 Fast Wide controller, 3.5-in. disk drive, CD-ROM drive, 8 mm 5/10 GB tape drive, graphics adapter Power GXT150M). This setup has 2 x 2 GB internal and 4 x 4 GB external hard disks, and was later upgraded with an additional 6 x 9 GB for a total of 74 GB, to "mirror" CrossFire*plus*Reactions and all other databases on this server for continuous operation.[38] The operating system, AIX 3.2.5, was upgraded in September 1996 to version 4.2 as a prerequisite for loading MDL's ISIS/Host 2.0.1. As the more than 100 RS/6000 computers at ETH Zürich are regarded as workstations, there is no operational support for such machines provided by the ETH Computer Center. This was the first Unix computer in the chemistry library and the first computer beyond the PC where the operating had to be done,[39] and library staff had to be trained on the job for this task. The server was delivered on November 15, 1994, and CrossFire was installed in the first half of April 1995. CrossFire was first made available[34] in the Electronic Library[40] of the ETH Chemistry Information Center (the chemistry library), which opened on June 1, 1996[41].

User administration is streamlined and uses electronic communication as much as possible to reduce administrative load on the library staff. All users must sign a license agreement, which stipulates that searches are permitted only in the context of their work at ETH, and that such searches cannot be executed on behalf of third parties outside ETH, nor search results given to such third parties. Taking a one-hour introduction to CrossFire is another prerequisite for getting an account. For research groups and users that do not access CrossFire often, we also offer group accounts. Such group accounts entail a security risk because the password must be known to several people, whereas responsibilities are not so clear. We enforced the usual password restrictions, and for the group accounts, a maximum password age of three months seemed advisable. In order to lessen the burden on our users, we decided to extend this to six months, however. We compensate for this by access control using the well-known software tcp_wrapper[42], which denies access by clients who are not calling from registered subnets at ETH Zürich or other universities that share our server[32].

Every research group participating in the CrossFire project has a contact person who is responsible for collecting account requests and informing users within her or his group about changes and problems. Lists with user names received via e-mail can be fed into Unix shell scripts to create and activate user accounts. Administration of user and account data is handled by an appropriate extension of the user file in our integrated electronic library system CLICAPS (Chemistry Library Information Control and Presentation System). This system was developed on the basis of Filemaker PRO networked on Macintosh PCs,[43] by automating as many routine operations as possible. For example, a script is used for printing out license agreements in duplicate bearing all the user and account data. In case of problems,[38] all contacts are informed via an e-mail mailing list, and they are expected to convey this information to their user community.

Communication with users is handled via two dedicated CrossFire WWW pages.[36] Solutions to problems that seem to occur frequently are offered via this communication route,[36] and users are urged to look at these pages before they request help from their colleagues or the library staff. When we noticed, for example, that many users were so fond of CrossFire that they used it even for author searches, we decided to compensate for that somewhat dangerous search technique by posting the author coverage of Beilstein for the respective time periods on this Web page and comparing it to that offered by Chemical Abstracts and the Science Citation Index.

The client software, Beilstein Commander for Windows or Macintosh, is distributed via the ETH file transfer protocol (FTP) server for licensed software, from where it is downloaded and installed within research groups by the aforementioned contact persons. Manuals are available in electronic form as Word 6.0 for Windows documents; a printout is available in the Electronic Library.

As the IBM RS/6000 is a dedicated database server, security measures beyond logging and basic auditing and accounting can be taken without impairing the legal users: all services known to carry security risks (*finger*, the UNIX *r* commands), or that are known to be used by hackers to gain information are either completely deactivated or restricted to privileged users. Client–server connections can be established either via *telnet* or *rexec*. The latter is the default set by Beilstein and was used by us until January 1996, because it is faster than *telnet* in establishing a connection, and there are no restrictions on the maximum number of concurrent users. With respect to some security problems,[44] and because sessions via *rexec* are not automatically logged by the operating system in the /var/adm/wtmp file, which is useful for statistics and access supervision, we switched to *telnet* mode for accessing the CrossFire server.

Acknowledgments

We gratefully acknowledge the Academic Programs provided by Beilstein, CAS, MDL, and STN. Without them, academic institutions such as ours would not be able to provide the amount of electronic information that is desirable and necessary for both education and research. I personally appreciate the support, which was beyond what was expected, given by J. Swienty–Busch, Barbara Kalumenos, R. Lohmeyer (Beilstein Information Systems GmbH), H. Kottmann, and G. Grethe (MDL Information Systems, Inc.). Comments from the joint beta testing conducted with my colleague S. A. Bizzozero were very helpful in preparing this manuscript.

References and Notes

1. Finch, A. F. *J. Chem. Inf. Comput. Sci.* **1986**, *26*, pp 17–22.
2. The completely nongraphical CRDS, including both the older Theilheimer volumes and *Journal of Synthetic Methods* produced by Derwent Ltd., is still offered by Questel–Orbit. Versions with structure search and display are THEIL (in-house for REACCS and ISIS from MDL Information Systems, Inc.; Vols. 1–35, 1946–1980), and *J. Synth. Meth.*, in-house as JSM (REACCS, ISIS; 1980–) or publicly as STN DJSM (1975–, available since December 1995).
3. Kasparek, S. V. *Computer Graphics and Chemical Structures*; Wiley: New York, 1990; Part Four. Kos, A. J.; Grethe, G. *Nachr. Chem. Tech. Lab.* **1987**, *35*, pp 586–594. McHale, P. J. *On-line Inf. [Proc. On-line Inf. Mtg.]* **1989**, *13*, pp 155–160. Grethe, G.; Moock, T. E. *J. Chem. Inf. Comput. Sci.* **1990**, *30*, pp 511–520. Christie, B.; Moock, T. In *Chemical Structures 2*; Warr, W. A., Ed.; Springer: Berlin, 1993; pp 469–483.
4. Zass, E.; Müller, S. Chimia **1986**, *40*, pp 38–50. Zass, E. In *Software Development in Chemistry 4*; Gasteiger, J., Ed.; *Proc. 4th Workshop Computers in Chemistry*; Springer: Berlin, 1990; pp 243–253.

5. Borkent, J. H.; Oukes, F.; Noordik, J. H. *J. Chem. Inf. Comput. Sci.* **1988**, 28, pp 148–150.
6. Chodosh, D. F.; Hill, J.; Shpilsky, L.; Mendelson, W. L. *Recl. Trav. Chim. Pays–Bas* **1992**, 111, pp 247–254.
7. Johnson, A. P. *Chem. Br.* **1985**, 21(1), pp 59–67. Hopkinson, G. A.; Cook, A. P.; Buchan, I. P.; Reynolds, A. E. In *Chemical Structures 2*; Warr, W. A., Ed.; Springer: Berlin, 1993; pp 459–468. Miller, T. M.; Boiten, J.-W.; Ott, M. A.; Noordik, J. H. *J. Chem. Inf. Comput. Sci.* **1994**, 34, pp 653–660.
8. Our first installations in winter 1984 included only 32,000 (Theilheimer and Organic Syntheses on REACCS version 6.0), 5000 (ORAC version 5.3), and 25,000 (SYNLIB version 2.2) reactions, respectively.
9. Blake, J. E.; Dana, R. C. *J. Chem. Inf. Comput. Sci.* **1990**, 30, pp 394–399. Langstaff, E. M.; Sobala, B. K.; Zahm, B. C. In *Chemical Information 2*; Collier, H. R., Ed.; *2nd Proc. Int. Conf. Montreux 1990*; Springer: Berlin, 1990; pp 17–24.
10. Parlow, A.; Weiske, C.; Gasteiger, J. *J. Chem. Inf. Comput. Sci.* **1990**, 30, pp 400–402. Gasteiger, J.; Weiske, C. *Nachr. Chem. Tech. Lab.* **1992**, 40, pp 1114–1120. ChemInform RX, produced by the Fachinformationszentrum Chemie GmbH, Berlin, is available in print and as a database with coverage starting in 1991, both publicly at STN, and as one of the most important in-house reaction databases via MDL's REACCS or ISIS.
11. In-house versions for MDL's REACCS of ChemInform and ChemReact had been available since 1991.
12. Schinzer, D. *Nachr. Chem. Tech. Lab.* **1993**, 41, pp 826–828. The InfoChem reaction database with about 390,000 reactions selected algorithmically (via machine reaction center recognition) from a pool of 2.5 million reactions is available as an in-house system under REACCS (*see* reference 14) or ISIS, or as public database STN ChemReact. Smaller subsets (plus a CD-ROM for displaying the entire pool of reactions) are available for PCs from Springer-Verlag [*see* Warr, W. A. *Database* **1994**, 17(3), pp 56–63].
13. Reaction databases in the figure are not discussed in the text or in references 2, 10, 12, and 14: CCR, Current Chemical Reactions (Institute for Scientific Information, Philadelphia, PA 19104; *see* Garfield, E. *Current Contents* **1987** (13), pp 3–7), in-house database for REACCS or ISIS (MDL Information Systems Inc., San Leandro, CA 94577); MOS, Methods in Organic Synthesis (Synopsys Scientific Systems, Leeds, U.K.); ORAC Core, reactions from the ORAC system[7] converted to REACCS and ISIS format by MDL. Besides the reaction databases covering general organic syntheses shown in Figures 7.1 and 7.2, there exist a number of in-house databases on special reaction types such as protecting groups, solid-phase syntheses, asymmetric syntheses, heterocycles, etc.
14. The ETH REACCS (MDL Information Systems, Inc., San Leandro, CA 94577) installation contained the following databases with a total of about 450,000 reactions: CLF (Current Literature File, 1983–1991), THEIL (Theilheimer Vols. 1–35, 1946–1980, *see* reference 2), ORGSYN (Organic Syntheses 1921–), CSM (*see* reference 24), and ICRDB (InfoChem Reaction Database, older in-house version of ChemReact[12] covering 1975–1987).
15. Grethe, G. In *Proc. 1995 Int. Chem. Inf. Conf., Nimes, 1995*; Collier, H. R., Ed.; Infonortics, Ltd.: Calne, U.K., 1995; p 40.

16. *See* Chapter 2 in this book and Luckenbach, R. *J. Chem. Inf. Comput. Sci.* **1981**, *21*, pp 82–83. Luckenbach, R.; Sunkel, J. *J. Chem. Inf. Comput. Sci.* **1989**, *29*, pp 271–278. Chrzastowski, T. E.; Blobaum, P.M.; Welshmer, M. A. *Serials Librarian* **1991**, *20*(4), pp 73–84.

17. Jochum, C. *World Pat. Inf.* **1987**, *9*, pp 147–151. Bucher, R.; Jochum, C. *IATUL Quart.* **1991**, *5*, 94–107. For AutoNom and Current Facts in Chemistry, *see* Chapters 10 and 4 in this book, respectively; for SANDRA, *see* Lawson, A. J. In *Graphics for Chemical Structures;* Warr, W. A., Ed.; ACS Symposium Series 341; American Chemical Society: Washington DC, 1987; pp 80–87. Wolman, Y. *J. Chem. Inf. Comput. Sci.* **1987**, *27*, p 144.

18. Although searching for implicit reaction information in the Beilstein database is possible in both its STN and KR DIALOG implementations, we almost exclusively used STN because of the synergies with CAS databases, the CAS Academic Program, and the convenient structure input via either Messenger commands (preferred by information specialists) or STN Express. For a comparison of both Beilstein database implementations, *see* Buntrock, R. E.; Palma, M. A. *Database* **1990**, *13*(6), pp 19–34.

19. *The Beilstein Online Database. Implementation, Content, and Retrieval;* Heller, S. R., Ed.; ACS Symposium Series 436; American Chemical Society: Washington DC, 1990; Chapter 7, pp 88–112.

20. The "filter" Preparation in the View menu, activated by a mouse click, reduces the information displayed and the number of reactions shown in the Field Availability display by eliminating preparations of other compounds *from* the title compound. Only reactions leading *to* the title compound, both those classified as Preparation (i.e., intentionally), as well as Chemical Behavior of other compounds leading to the formation of the title compound (though not necessary with a preparative intent), are shown. Nongraphical Reactions are not affected by this filter. However, the results are not identical to those of the former (CrossFire version 2) data field Preparation, as reactions leading to the title compound, but not executed for preparative reasons such as kinetics, are also included. These may also be of preparative interest, but not necessarily so; excluding them is possible by searching only for "Reaction Classification: Preparation" (a slow process).

21. Jochum, C. *J. Chem. Inf. Comput. Sci.* **1994**, 34, pp 71–73. Hicks, M. G. *J. Chem. Inf. Comput. Sci.* **1990**, *30*, pp 352–359.

22. Examples 1, 3, 4, 7, and 8 are recent problems, while the remaining six examples where older searches that we also tried in CrossFire*plus*Reactions and updated in the other databases discussed. For discussions of examples 2, 6, and (in part) 11 at the "state of the art" in 1990, *see* Zass, E. *J. Chem. Inf. Comput. Sci.* **1990**, *30*, pp 360–372.

23. I thank Beilstein Information Systems GmbH for permission to run this and other queries on its server in the United States before we were able to install our own version of CrossFire*plus*Reactions.

24. As we had only CSM (Current Synthetic Methodology, a manually selected subset of ChemInform RX with ca. 27,000 reactions, 1992–1995) available in-house on REACCS (*see* reference 14), we used STN ChemInform RX in the examples discussed here. It is important to notice in this context that everything that was retrieved here in STN ChemInform RX should also be retrieved by the in-house version with a better user interface.

25. This feature is of course no substitute for large public (STN CSCHEM/ CSCORP, Chemcats) or in-house (Available Chemicals Directory (ACD) for MDL's REACCS or ISIS) catalog databases.

26. I found STN SmartSelect for reagents or solvents in CASREACT too slow and too expensive, and establishing the context between information in CASRE-ACT/Registry/CA too unwieldy to make it a routine operation for end users.

27. In ISIS, "extracted" information can be exported into an Excel spreadsheet for viewing and further processing.

28. This of course works also for a Markush-type substructure search for structures or reactions: when all atoms in Figure 7.10 had the maximum number of free sites, 124 reactions were retrieved, or 94 publications when searched for with the identical query in the Citations context. Twenty-one of these were published before 1967, including the "classical" work of Bredt et al. on camphor in the 1930s as well as a paper by Baeyer and Villiger from 1898.

29. Beilstein has not been covering patents after 1980; before that time, patent coverage can be considered useful for the time period 1960–1979 (21% of all compound records in CrossFire contained at least one patent), and good for the period 1800–1949 (98.5%) and 1950–1959 (94%). CASREACT has covered reactions from patents since 1991.

30. This method is much less expensive in the public databases CASREACT and ChemInform RX than an exact structure search, which unfortunately costs the same as a reaction substructure search.

31. The amount of time necessary to make detailed comparisons of search results (in contrast to comparing just lists of references retrieved) usually prevents the examination of enough examples to arrive at reliable general conclusions.

32. Examples for shared use of CrossFire are (1) a consortium of midwestern American universities (Committee on Institutional Cooperation (CIC), *see* Chapter 8 in this book; now as the Minerva consortium, it is open to all U.S. universities and colleges), (2) a group of five universities in the German state of Baden-Württemberg, and (3) MIDAS (Manchester Datasets and Associated Services) universities in the U.K. Our server at ETH has also been used by the University of Basel, with the Universities of Berne, Fribourg, Neuchâtel, Lausanne, and Geneva joining in 1997. The first of such academic consortia for using reaction databases and other databases and software, the CAOS/CAMM Center in Nijmwegen, Netherlands, recently celebrated its 10th anniversary (*see* Noordik, J. H. In *Chemical Structures;* Warr, W. A., Ed.; Springer: Berlin, 1988; pp 367–370. Thiers, A. H. M.; Leunissen, J. A. M.; Miller, T. M.; Schaftenaar, G.; Noordik, J. H. *J. Chem. Inf. Comput. Sci.* **1990**, *33*, pp 858–862.

33. Lawson, A. J.; Swienty–Busch, F. *On-line Inf.* **1993**, *17*, pp 187–194. Zirz, C.; Sendelbach, J.; Zielesny, A. In *On-line und darüber hinaus . . . Tendenzen der Informationsvermittlung;* Neubauer, W.; Schmidt, R., Eds.; *Proc. 17 On-line-Tagung DGD; Dtsch. Ges. Dok.*: Frankfurt/Main, 1995; pp 247–257. Zass, E.; Donner, W.; Sendelbach, J.; Zirz, C. In *Proc. 1995 Int. Chem. Inf. Conf., Nimes, 1995;* Collier, H. R., Ed.; Infonortics Ltd.: Calne, U.K., 1995; pp 138–144.

34. Until April 1996, CrossFire was available to users only at three Windows PCs in our Electronic Library[40] for two reasons. First, we wanted to observe the users and obtain direct feedback about their problems with CrossFire, which was easy, as the Electronic Library is supervised all the time, but it was difficult with use at the bench. Second, the organic chemists, the major users of CrossFire, almost exclusively have Macintosh PCs, for which client software became available only at that time.

35. Zass, E. In *Chemical Information*; Collier, H. R., Ed.; *Proc. Int. Conf. Montreux 1989*; Springer: Berlin, 1989; pp 55–62. Zass, E. *Mitteilungsbl. Ges. Dtsch. Chem. Fachgruppe Chem. Inf.* **1994** (29), pp 13–32. For the current chemical information instruction program in the ETH chemistry library, *see* http://www.chem.ethz.ch/chembib/Ausbildung.html.

36. CrossFire WWW pages: one offers a general description, tips, tricks, and supplementary information (http://www.chem.ethz.ch/chembib/Xfire.html), the other gives access conditions and installation instructions (http://www.chem.ethz.ch/chembib/Xfire2.html). These WWW pages are needed because users cannot be expected to read voluminous manuals; so like surfing on the World Wide Web, we choose this avenue for transmitting important information to them.

37. With reaction searching in CASREACT implemented recently in CAS SciFinder 2.0, this has been improved significantly. For academic institutions, however, SciFinder (Williams, J. *On-line* **1995**, *19*(4), pp 60–66. Williams, J. In *Proc. 1995 Int. Chem. Inf. Conf., Nimes, 1995;* Collier, H. R., Ed.; Infonortics, Ltd.: Calne, U.K., 1995; pp 145–159) is hardly affordable at present, and it is certainly not competitive with CrossFire (with or without reactions) at current rates and charging options.

38. CrossFire was down for about a week at the end of February 1996 because of a defective power supply in one of the 4-GB hard disks.

39. REACCS, SYNLIB, and the now discontinued ORAC were loaded by us, but the entire operating and user administration has been handled by the ETH Computer Center.

40. The Electronic Library is a permanently staffed room within the ETH chemistry library (opening times are the same as those of the library, less Friday afternoons and one hour each in the mornings and evenings), dedicated to electronic sources. In 1996, the Electronic Library contained nine PCs (one DOS, five Windows for Workgroups 3.11, and three Macintosh), which served as publicly accessible terminals to in-house databases, the Internet, locally loaded software (e.g., SANDRA, AutoNom), and 18 CD-ROM databases on four Pioneer 18-CD-ROM and three Nakamishi 7-CD-ROM changers. *See* http://www.chem.ethz.ch/chembib/El_Bibl.html and Zass, E. In *On-line und darüber hinaus . . . Tendenzen der Informationsvermittlung;* Neubauer, W.; Schmidt, R., Eds.; *Proc. 17 On-line-Tagung DGD; Dtsch. Ges. Dok.:* Frankfurt/Main, 1995; pp 171–187.

41. This delay was not caused by CrossFire or by difficulties operating the new Unix server, but by the many problems and the often unsatisfactory producer's support incurred when installing the different CD-ROMs in the Electronic Library. At that time, all of the hardware and software installations, including those needed for the operation of CrossFire, were done by one person who also had other duties, including most of the structure-based searches in public databases in the ETH Chemistry Information Center (4–10 hours per week).

42. Access control with tcp_wrapper may be restricted down to individual IP addresses, but for the sake of user flexibility within acceptable limits, and to lessen the administrative effort involved, we decided to stay at the subnet level for access control.

43. Suter, D. P. In *On-line und darüber hinaus . . . Tendenzen der Informationsvermittlung;* Neubauer, W.; Schmidt, R., Eds.; *Proc. 17 On-line-Tagung DGD; Dtsch. Ges. Dok.*: Frankfurt/Main, 1995; pp 31–59.

44. Some networks filter all Unix r commands, including *rexec,* for security reasons, and recently (April 1996), there was an organized attack on computers at the ETH and numerous other sites in Switzerland using *rexec.*

Chapter 8

Beilstein's CrossFire: A Milestone in Chemical Information and Interlibrary Cooperation in Academia

Ken Rouse and Roger Beckman

The use of CrossFire in a large university consortium in the midwestern United States is described, including the experiences of a substantial number of academic end users at these universities. Accessibility of Beilstein via CrossFire is compared with the accessibility of other databases of chemical information used in academia.

In the short time since it was announced in 1994, Beilstein's CrossFire, a new, computerized rendering of the venerable *Beilstein Handbuch der Organischen Chemie*, has generated excitement in every corner of the chemical information world, earned awards for the prime movers behind its creation,[1] and reestablished Beilstein as a force to be reckoned with in the academic market. A large group of academic users in the United States gained early access to CrossFire via a consortium agreement brokered by the Committee for Institutional Cooperation (CIC), an organization comprised of 12 midwestern universities that is sometimes referred to as the academic equivalent of the "Big Ten" athletic conference.[2] The response to CrossFire among this group has been overwhelmingly positive. Chemistry librarians, who for the most part promoted CrossFire to chemists at their respective institutions, were surprised and gratified to see how quickly organic chemistry faculty and graduate students embraced Cross-Fire and began incorporating it into their daily research routines. They were even more surprised to see how quickly a sophisticated research tool such as CrossFire made its way into the undergraduate curriculum on some campuses.

This chapter examines the CrossFire phenomenon from the perspective of two CIC universities: Indiana University and the University of Wisconsin. First, the developments that led to Beilstein's decline in academia are reviewed. Then the major reasons for Beilstein's promising renaissance with CrossFire are discussed:

1. the use of cutting-edge computer software and hardware that has transformed a work that even experienced researchers found daunting into a resource that can be used—with a modest investment of time—by anyone needing information about organic compounds,
2. Beilstein's decision to offer CrossFire for a fixed, annual fee, which experience has shown to be the sine qua non of end-user access, and
3. innovative pricing strategies, as exemplified by the CIC consortial agreement.

Beilstein in Academia Before CrossFire

To appreciate what Beilstein has accomplished with CrossFire, one must be aware of Beilstein's precarious situation in the U.S. academic market before its arrival. Considered for generations to be an indispensable reference in any chemical research collection, by the 1980s a variety of developments had conspired to make the *Handbuch* an ever-rarer item in academic libraries.

The information explosion, for one, so overwhelmed the famously rigorous Beilstein system that the Institute's more than a century-old goal of carefully documenting the structure and important data for every organic molecule appeared more and more to be a mission impossible. By the 1980s, 10 or more physical volumes were required to cover what had once fit in one, and the *Handbuch* (still written in German at that time) was lagging from 15–35 years behind the current literature. This lack of currency was undoubtedly a major obstacle to the modern use of Beilstein. In the experience of the authors, it has been very difficult to motivate researchers to use a work as complex as Beilstein when the references retrieved are decades out-of-date.

Most threatening to Beilstein's existence in academia was (and still is) what librarians have come to call the "serials crisis". Beginning in the 1960s, commercial publishers began to exploit—and many would say "abet"—the information explosion by offering researchers new and usually less expensive outlets for their publications. Most of these titles were seductively affordable when they first appeared, but as the literature expanded relentlessly and prices increased by as much as 30% in some years, a vicious price spiral ensued: cancellations begat higher prices, which in turn produced more cancellations and higher prices. By the 1990s, the $20 journal of the 1960s had become the $6000—or even $15,000—budget breaker.[3] As research collections at even the best-funded universities were downsized, academic librarians began preaching the virtues of "access rather than ownership", by which was meant the provision of photocopies of individual articles from expensive journals in lieu of paying for subscriptions. Beilstein itself was also caught up in the journal price spiral. By the early 1990s the *Beilstein Handbook* carried a price tag in excess of $30,000, and it became a frequent target for cost savings.

The rise of English as the international language of science and the decline in the study of German are often cited as a trend that contributed to the decline of Beilstein in academia, by rendering it less accessible to the growing numbers of chemistry researchers and students who were no longer required to read German. *Beilstein's Handbuch der Organischen Chemie* itself bowed to the trend and switched to English in 1984 with volume 17/1 of the 5th Supplementary Series, and is now known as *Beilstein's Handbook of Organic Chemistry*. Although the Handbook's being written in German was certainly an obstacle for some users, it is likely that the trend toward English in science did more harm to Beilstein in a more indirect sense, that is, as one aspect of the information explosion and the serials crisis. Journal publishers have found in the increasing flood of research from around the world—now written in English and therefore marketable—a seemingly inexhaustible supply of articles with which to fill new volumes of high-priced journals.

Computerization was ultimately to be the salvation of the Beilstein enterprise in academia, but until CrossFire appeared, it was another trend that undermined the viability of the handbook. Computerization, in the first place, put additional stress on library budgets by diverting large amounts of money away from print collections and into computer products, infrastructure, and services. Second, as database searching of *Chemical Abstracts* and other files became common in the 1970s, the fact that computerization of Beilstein was still in the future reinforced the notion in many minds that Beilstein was old fashioned and expendable. By the time Beilstein Online was launched on the STN network in 1988 and shortly thereafter on DIALOG, it encountered stiff competition from Chemical Abstracts Service (CAS) and other products. A survey conducted in 1993 showed that Beilstein Online had failed to gain wide acceptance at universities, despite the availability of liberal academic discounts and round-the-clock access.[4] At the University of Wisconsin, Beilstein Online was used almost exclusively for quick reference searches paid for by the library. When researchers had money to spend for online searching, they were inclined to remain with the familiar, more up-to-date offerings of CAS and relied on the references in newer papers to lead them to the most important older sources.

In 1991 Beilstein introduced a CD-ROM product, Beilstein Current Facts (CF). CF offered current Beilstein data (from 1990) on quarterly updated disks that eventually contained one year's worth of data. Described as a "benchtop tool" for the end user, CF was very impressive in its own right and foreshadowed many of CrossFire's strengths. In fact, the Windows version of CF, which appeared about a year before CrossFire, was basically version 1.0 of the CrossFire client. It was easy to use, even in its DOS version, and—most important—its faster search algorithm made structure searches possible on a desktop computer for the first time.[5] CF attracted some enthusiastic users at a number of universities (among them Indiana University and the University of Wisconsin), and those who purchased it got a valuable preview of CrossFire. But the built-in limitations of the CD format—slowness and limited data coverage—probably explain why it has not been a bigger success in the academic market.[6]

By the early 1990s it appeared to many academic observers that Beilstein might be on the road to oblivion. Its computer offerings were being largely ignored by academic customers, and the "serials crisis" had entered an end-game phase in which some libraries began to cancel all offerings from pricier publishers. Under these circumstances, the temptation to avoid a year of cancellation-induced stress by throwing a *Beilstein Handbook* overboard became ever more irresistible. The *Beilstein Handbook* still had its defenders, among them not a few librarians who protested the sacrifice of a work of proven, long-term value for the short-term objective of sparing a handful of expensive journals until the next bloodletting. Certainly, if "access rather than ownership" was to be the watchword of librarians on their way to the brave new world of virtual librarianship, it did not make much sense to cancel a work that enabled researchers to extract valuable data from the mounting glut of frequently never-used information. Despite these and other arguments, *Beilstein Handbook* subscriptions continued their free fall. By the eve of CrossFire's appearance, *Handbook* subscriptions had declined from over 3800 in 1958 to 3300 subscriptions in 1965 to about 295 in 1994, and in 1996 to only some 260.[7]

Reception and Use of CrossFire

To find an event that generated as much interest in academic chemistry libraries as the appearance of CrossFire, one would have to go back to 1984, when the announcement of the original "Academic Program" by the CAS raised hopes that a new era of widespread end-user access to CAS files was dawning. CrossFire has been available in the United States only since the spring of 1995, but judging from the number and enthusiasm of users, it appears that Beilstein may have found in CrossFire the means to recover much of the ground it has lost since the 1960s.

Academic library users these days must cope with a confusing array of networked (or non-networked) databases with different interfaces that seem to change about the time one has grown comfortable with them. In this environment, new offerings from the library are not always welcomed by users with unlimited delight, and library memos extolling the virtues of a new product not uncommonly end up unread in the wastebasket. CrossFire was a major exception. The authors have yet to interview a user who, once persuaded to have a look at CrossFire, was not immediately impressed if not awed by its capabilities. The following sample of user comments from Indiana University (IU) and the University of Wisconsin (UW) make it clear that CrossFire was not perceived as just another database with useful information, but as a major new resource that was changing the way researchers work:

- "I estimate this is saving me about 5 hours per week." (UW organic chemistry graduate student.)
- "This is great. With [Beilstein] Commander you don't have to be an expert in organic synthesis to do organic chemistry." (IU organic chemistry graduate student.)
- "With CrossFire my graduate students are actually finding the compounds they are looking for." (UW organic chemistry professor.)
- "I don't know how I ever got through graduate school without this." (UW organic chemistry professor.)
- "We don't run a reaction without checking CrossFire first." (UW pharmacy graduate student.)

Graduate students and faculty at Indiana and Wisconsin were dutifully informed about CrossFire via announcements in departmental and library newsletters, but the most effective way of spreading the word about CrossFire at both universities proved to be brief, one-on-one demonstrations that were offered by library staff. Once a few graduate students were initiated, word spread quickly through the organic chemistry laboratories, with one student introducing others in his or her research group, or perhaps motivating others to ask library staff for a demonstration. At neither school has there been much demand for more formal, in-depth, group training sessions, although at Indiana new graduate students as a group get a brief look at CrossFire as part of an introduction to library resources. At Wisconsin only a few group training sessions have been given so far, and these were largely for the benefit of library staff in other science libraries, an exception being a demonstration requested by the Chemical Engineering Department.

Not surprisingly, organic chemistry graduate students form the largest group of CrossFire users, but at least at Wisconsin, physical chemists, chemical engineers, and researchers in other chemistry-related fields are discovering CrossFire and use it to locate data on compounds of interest. The organic chemistry graduate students, in most cases, take advantage of CrossFire's structure searching features to gain easy access to preparation data. A summary observation by a graduate student at Indiana is perhaps typical:

It's fast, cheap, and convenient. More so than CAS Online because finding preparation is so easy. I also use it to find out characteristics of a compound that will help me design experiments in the lab. Such things as NMR data, boiling point, if it has been crystallized. I want to know if it has been made and if not what has been made that's similar.

The sudden popularity of CrossFire caused library staff to scramble to provide adequate access. Wisconsin began with a single CrossFire workstation, but when students complained of long waits if they did not come in late at night, the library pushed ahead with plans to bring Cross-Fire up on most of the computers on the campus network, thereby providing access from six computers in the Chemistry Library and from many more in other science libraries. The Indiana experience was similar, with access beginning with a single staff computer and then expanding to several public machines as demand increased. Indiana was also able to load CrossFire on eight networked machines located in an instructional computer laboratory located adjacent to the library.

Beilstein's new Commander software, which was introduced along with reaction searching in the fall of 1995, increased the capabilities of CrossFire significantly, but the more powerful interface brought with it a number of technical and access problems, some of which are still unresolved at some sites. In the University of Wisconsin's networked environment, which employs Novell software and Windows 3.1, a number of machines currently are unable to print structural diagrams or take advantage of the "print-to-file" option that allows for limited downloading: that is, a PostScript file containing a structural diagram and accompanying factual information. (CrossFire's limited downloading capabilities are seen by some observers as one of its major drawbacks.) There are some indications that an upgrade to Windows 95 may solve the problems that arose with Commander's greater complexity, but such an upgrade is not imminent. At Indiana, the higher processing speed and memory requirements (a computer with a 486 CPU and 8 MB of main memory) of Commander ended access from several 386 CPU public computers and caused conflicts with the configuration of the Windows NT 3.51 workstations. This conflict has been solved by upgrading the workstations to Windows NT 4.0.

These and other problems, such as the occasional and usually inexplicable computer crash that requires a reboot, are certainly frustrating, but they have not been enough to dampen the enthusiasm of CrossFire users. In the last few months six organic chemistry faculty members at UW and four at IU have installed CrossFire on their office computers. Others are eager to do so, but, being primarily Macintosh users, they have elected to await the availability of the new Macintosh client. The lack of a Macintosh client, as has been documented in an earlier study,[8] has certainly been a significant impediment to wider faculty use of CrossFire, especially at Indiana, where the members of the organic chemistry faculty are primarily Macintosh users. Beilstein announced a Macintosh client in the fall of 1995, but persistent problems have prevented its being deployed at Wisconsin and hence at the other CIC institutions. A new, improved Macintosh client was successfully tested at Wisconsin (May 1996) and is now available across the CIC. It is being used successfully in several laboratories and in the library at Indiana University.

CrossFire had a nearly immediate impact on graduate and faculty research, but it has not taken chemistry faculty long to realize the potential of CrossFire as a teaching tool. A few months after CrossFire became available to the CIC, there were reports that CrossFire was being installed in chemistry department computer laboratories to provide better access for students, and there were some concerns (which so far have proved groundless) about overtaxing the Beilstein server. The head of the Physical Sciences Library at Pennsylvania State University recently made the following observation: "The faculty response to this has been great; but the students . . . WOW!! Some of the faculty and I have been integrating it into our [undergraduate] classes. . . . Many of the students have commented on how much easier their course work and research efforts are because of this."[9]

At UW, CrossFire now plays a major role in the undergraduate laboratory course in which students are introduced to library research methods. Moreover, several professors who teach a distance education course in organic chemistry for students at two-year campuses around the state plan to use CrossFire in their course. A similar trend is also developing at IU. Undergraduate chemistry students at IU are introduced to CrossFire through a one-credit, required course in chemical information that most take when they are juniors or seniors. Several IU faculty members also plan to incorporate CrossFire into upper level organic chemistry laboratory courses in the coming year. Encouraged by the ease with which students take to CrossFire, the University of Wisconsin and the University of Illinois at Chicago are even planning a pilot project to test the use of CrossFire as a teaching tool in several high schools.

The remarkably enthusiastic reception afforded CrossFire in academia is an interesting story in its own right, but it is becoming increasingly clear that CrossFire is also beginning to make significant inroads in the industrial market. Competitors have taken note. That *Chemical Abstracts* now views Beilstein as a serious competitor is suggested by its recent decision to deny Beilstein access to CAS Registry Numbers for compounds added to the Beilstein database after 1994.[10] Registry Numbers included in CrossFire after that date will be those that appear in the original publications.[11] This means that CrossFire users must use caution when searching on Chemical Abstracts (CA) Registry Numbers higher than 159909–17–8.[12] Beilstein of course assigns its own registry numbers, the BRNs, and has registered older compounds that are not in the CA Registry File. If the new Beilstein enterprise continues to prosper, Beilstein's and CAS's customers may well begin to wonder whether their interests would not be better served if there were only one Registry Number that could be used to search for compounds in both databases.

The CrossFire Difference

The existence of CrossFire is unthinkable, of course, without the decade of prior effort that went into the creation of the Beilstein Online database and eventually Beilstein Current Facts. Several major problems contributing to the handbook's decline were thus confronted before CrossFire's appearance. Most important, the currency of Beilstein's coverage of the literature was vastly improved, albeit at the expense of breadth of coverage. Both CrossFire and Beilstein Online are now updated quarterly and lag approximately six months behind the literature. To achieve this currency, coverage is limited to 120 of the most important organic chemistry publications. Beilstein Online also dealt with the language barrier problem by adopting English as the search language, although German is still very much in evidence in text fields. As important as the resolution of the currency and language problems was, it is clear that the technical innovations introduced with CrossFire contributed significantly to its popularity with users.

CrossFire employs IBM RISC-based client–server technology. Its faster search algorithm, user-friendly Microsoft Windows graphical interface, and hyperlink technology combine to offer unparalleled access to the structures and related data of more than 7.5 million organic compounds. The new Commander version of CrossFire adds the largest currently available reaction database (5 million reactions) and various software improvements, the most impressive being a major expansion of the hyperlink feature. In Commander, substances, reactions, and citations are hyperlinked so that the user may move quickly and effortlessly among the millions of structures, reactions, or citations by clicking on the appropriate hyperlinked number assigned to each. For example, one can bring up the preparation data for a compound and then hyperlink to the various reactants or products and study their structures and properties. Or one can click on a citation identifier associated with that preparation and within seconds be browsing through the structures and reactions contained in the article.

Users frequently comment on how easy CrossFire is to use, and this is certainly one of its major advantages. The briefest of introductions to CrossFire's basic features is frequently sufficient to enable students—graduates or undergraduates—to begin using it to find in minutes information that might have taken hours or days to find in other ways. On the other hand, not everything about CrossFire is easy and apparent. The authors have found that many graduate students, although using CrossFire very effectively for their purposes, are just beginning to mine its potential. Like Beilstein Online, CrossFire employs hundreds of different fields and offers many different search options: exact structure, partial structure, Markush structure, formula, names, numerical data, etc. Various "hit sets", moreover, can be logically intersected in a variety of ways. For example, from the results of a partial structure search, one can extract those compounds that have desirable physical properties, such as a particular melting point range.

End Users, CDs, and the Fixed-Price Factor

CrossFire's fast, friendly interface was a very important factor behind its enthusiastic reception by academic end users, but of more fundamental importance was Beilstein's decision to offer CrossFire to all customers for a fixed price. University subscribers pay a single annual fee that includes the handbook and the CrossFire client and database. For end users to be enthusiastic about a database, they must first have convenient, affordable access to it. It was clear to Beilstein from the outset that CrossFire was unlikely to find its way into the hands of many end users if it were burdened with the pay-as-you-go approach of traditional online searching, with its open-ended costs and "taxi meter stress".[13] Beilstein's offer of its new database on an in-house, fixed-price basis, combined with the absence of any such offer from Chemical Abstracts Service for any of its databases,[14] goes a long way toward explaining CrossFire's early success in academia.[15]

In the past year, there has been considerable discussion of the "desktop" revolution, by which is meant the increasing availability of chemical information from the desktop computers of chemical researchers.[16] From the perspective of many academic libraries, however, the desktop revolution is only the latest development in an end-user revolution that began to gain momentum a decade ago. As the online movement progressed, it became increasingly evident that traditional online searching, which was typically mediated by librarian searchers, would never be able to meet the growing need of large numbers of academic users for access to computerized information. The solution was to allow users to do their own searching, but this would be impossible if costs could not be controlled. In response to this need, a mix of CD products and leased or purchased databases, all available for a fixed annual fee, began to establish themselves in many libraries in the mid-1980s. The sciences frequently led the way in this trend. By the early 1990s, at least at large research universities such as UW and IU, computerized indexes covering most disciplines in the sciences became available to end users on public access library terminals. As knowledge of CD networks progressed and campus-wide licenses were negotiated, some began to turn up on the desktop computers in faculty laboratories and offices. At UW, for example, Science Citation Index has been available in faculty offices for several years.

A major exception to this trend was chemistry. Chemical Abstracts Service, producers of *Chemical Abstracts*, the largest and most comprehensive index of chemical literature and allied disciplines, has been a pioneer in the computerization of chemical information, but until quite recently

CAS has had only a modest CD program.[17] Given the enormous size of the CAS Online files, CAS's hesitancy about using CDs to provide end users with fixed-price access is understandable. Less comprehensible, however, at least in the view of many academic users, was CAS's inability to find a way to give academic users what many have requested for many years: affordable, flat-rate access to the CAS Online files. CAS offered such an approach, with liberal access hours, with their first Academic Program in 1984, but within six months they retreated from the flat-rate/liberal access concept. The new Academic Program offered a 90% discount to academic users who accessed the system during evening and weekend hours. The discount was reduced to 80% in 1989, and over the years additional charges have been added. Moreover, as CAS added valuable new files such as CASREACT, a chemical reaction file, no discounted access was provided for academics, which had the effect of reserving its use for larger industrial customers. The same may be said of CAS's latest product, SciFinder, with which CAS hopes to serve the growing need for information from the desktop.

With the exception of those universities able to provide large subsidies, the lack of a fixed-price option has proven to be an insurmountable obstacle in the way of academic end-user access to *Chemical Abstracts* files. Judging from various surveys,[18] UW is typical of larger universities that did not subsidize CAS searching. After CAS abandoned its brief experiment with a flat-rate approach for the Academic Program, the UW Chemistry Department began charging back the cost of online searches to individual professors' accounts. The relatively generous but still limiting 90% discount program attracted a fair number of end users for several years, but a downward spiral in the number of both end-user and mediated searches commenced as costs increased and fixed-cost CD products presented an alternative. End-user searches averaged 20 per month in 1988–1989, 15 per month in 1991, 10 in 1992, and only 4.3 in 1993.

The problems awaiting those universities that attempt to subsidize CAS Online searching were illustrated in a recent article that describes the Indiana experience.[8] IU has a long tradition of strong support for chemical information and has been a national leader in chemical information education. In the 1970s, for example, IU was one of the few universities to lease CA computer tapes. With the appearance of the CAS Academic Program in 1984, IU elected to provide a total subsidy for CAS searching for all faculty, graduate students, and staff. Between 1984 and 1994, hundreds of end users performed more than 3000 searches per year. These numbers declined abruptly to an annual rate of approximately 280 when uncontrolled costs forced IU to end the subsidy and a charge-back system was introduced.[8]

The academic users who stopped searching CAS Online at UW, IU, or elsewhere as costs escalated did not, for the most part, return to the index tables in their libraries to renew their struggle with the 350 linear feet of printed CA indexes. They turned instead to other computerized indexes. When the CAS Online subsidy ended at IU, for example, the use of Science Citation Index on CD increased dramatically. Thus, it is not only the *Beilstein Handbook* that has been abandoned by many academic users of the last few decades, but printed CA as well. At the University of Wisconsin Chemistry Library in the 1970s, there were frequent complaints from faculty or graduate students about the abstract tables being appropriated by undergraduates for study purposes. In the 1990s, graduate students are more likely to be competing for a seat at a library workstation, where they will be searching Science Citation Index, Analytical Abstracts, Wilson Indexes, or some other computerized index available on an annual payment basis.

These indexes are not without their advantages, but they lack the breadth and power of indexes designed specifically for chemists. Most important, they lack structure searching capability. Structure searching is a uniquely powerful feature that provides chemists with information that they have great difficulty finding in any other way, if they are able to find it at all. The importance of structure searching is evidenced by the critical role it played in the long dispute, now fortunately settled, between CAS and DIALOG Information Services (now part of Knight-Ridder Information) over access to CAS's data. Initial skirmishes began during the early years of CAS Online when CAS introduced abstract searching but decided to withhold the abstracts from its major competitor. But the "last straw" that provoked a bitter, three-year court battle (1990–1993) was apparently CAS's refusal to share what may be its most precious resource: the connection tables that enable structure searching.[19]

Structure searches are relatively costly, even at academic rates, with a typical search on CAS Online involving one substructure and the display of 50 hits costing at least $30 (whereas the nondiscounted rate is about $100). Consequently, without a flat-rate option or a large subsidy, end-user access is all but out of the question. In the 11 years since the University of Wisconsin has participated in the CAS Academic Program, only a handful of graduate students have accepted the standing offer of training in structure searching, despite the availability of STN's excellent search software (STN Express), which makes basic structure searching relatively easy. Instead, structure searches were performed by the chemistry librarian. To be the sole keeper and dispenser of such a valuable resource is not without its appeal, but it cannot be compared with the satisfaction of seeing this powerful tool in the hands of hundreds of academic end users, which is now possible with CrossFire.

The CIC Arrangement

With a friendly, intuitive interface and subscription-based pricing, Cross-Fire is now well positioned to become the first major chemical database widely available to academic end users. However, on campuses where serial reduction exercises have become as much a part of the fall season as Big Ten football games, a product as expensive as CrossFire could have been a tough sell without the attractive pricing made possible by the CIC arrangement.

The CIC contract came about in the following manner. When Cross-Fire was announced in the spring of 1994 and offered to academic Handbook subscribers on relatively attractive terms (an additional $4500), a number of chemistry librarians in the CIC group set about making plans to make it available on their campuses. At the University of Wisconsin, these efforts were made easier when the chemistry and pharmacy departments quickly agreed to pay some of the hardware costs. When the central library administration was approached for additional funding, the Director of Libraries recognized in CrossFire an ideal candidate for a database that might be shared among CIC libraries via the Internet, that is, a large database with relatively high hardware and maintenance costs that would result in significant savings, but which was also sufficiently specialized so that usage would not be high enough to cause problems. The Associate Director for Automation at UW had been exploring just such a possibility with other CIC automation leaders when the CrossFire opportunity arose. In June 1994, the Director of Libraries at UW proposed that UW host the database on behalf of the CIC. The contract that eventually emerged from discussions between the CIC and Beilstein included all CIC universities with the exception of the University of Iowa. Two members of the CIC group that had canceled Beilstein were given several months of free-trial access as an inducement to resubscribe, and eventually agreed to do so. Three non-CIC universities that had expressed an interest were also admitted to the consortium: Wayne State University, the University of Cincinnati, and Iowa State University.

In the end, the CIC contract proved to be of great value for all the parties involved, although there were significant delays that tried the patience of some of the librarians, particularly those who were in a stronger position with respect to computer infrastructure and who might have had speedier access to CrossFire had they decided to go it alone. UW had hoped to have CrossFire available for all CIC libraries by the end of 1994, but for a variety of reasons it did not appear until March 1995. On the other hand, some universities might have had to forgo CrossFire had it not been for the CIC contract. From a fiscal point of view, the CIC CrossFire arrangement was the kind of opportunity that previously existed only in the fantasies of library directors during the early years of computerization, that is, a computer product that actually saved money. The 15 libraries that participated received a print subscription to the Handbook and unlimited access to the database for about $27,600. This was less than a print subscription alone had cost in the preceding year. But this was only part of the savings. Because the database could be mounted on a single server, each institution also saved the cost of installation and maintenance.

For the Beilstein organization, which is well aware of the importance of getting its new database into the hands of chemists while they are being trained, the CIC contract provided them with much quicker access to a significant part of the U.S. academic market than would have been possible otherwise. Whether for economic or technical reasons, some of the CIC universities were not prepared to mount the database locally, and at the time Beilstein was not ready to offer them access from its own server. Another obvious benefit for Beilstein was that instead of having to discuss installation and maintenance problems with computer staff at 15 institutions, they had to deal only with one person at the University of Wisconsin, who devised efficient procedures for distributing the client and responding to problems. The Beilstein client software was made available on the World Wide Web (WWW), with access controlled by use of Internet protocol (IP) addresses. Good communication among the CIC CrossFire participants was assured by library staff at the University of Illinois at Chicago, who created a CIC-BEIL listserv.

CIC members also provided a valuable assist with documentation. The first CrossFire manuals were impracticably long and contained a number of errors. The new Commander manuals are considerably better and include a tutorial. What was lacking when CrossFire first appeared was a brief guide that explained the essentials to beginners. Such a guide was provided by the chemistry librarian at the University of Illinois at Chicago and was made available from the library's WWW access, thereby sparing others the labor of reinventing the wheel across the CIC.[20] A subsequent collaboration with the chemistry librarian at the University of Chicago resulted in an HTML version of the same guide.[21]

The CIC organization itself has also benefited significantly from its involvement with CrossFire. In the years since it was established in 1958, the CIC has had some notable successes in promoting cooperation among member libraries, particularly with respect to interlibrary borrowing. In recent times, the networking of computerized information has enabled the CIC to set its sights much higher. Several months prior to the arrival of CrossFire, the CIC announced its Virtual Electronic Library project, or VEL, which was inspired by the following vision:

> A megalibrary as accessible to the students, faculty, and staff at each of the 12 major research universities of the CIC as their local libraries. The cornerstone of the project is the implementation of linked-system technologies connecting the online library catalogs of the CIC universities.[22]

Another component of the CIC's vision of the electronic future, as we have seen, was the sharing of databases among the various campuses. Before CrossFire, however, database sharing, as well as other aspects of

VEL, were "works in progress". The CIC's successful deployment of the CrossFire database—the first consortial license agreement of its kind—gave the entire VEL effort a significant boost early in its development. Since CrossFire, the CIC has been able to arrange consortial agreements for other databases, among them science databases such as MATHSCI and NTIS.

Implications for the Future

The new Beilstein organization has been moving quickly to exploit its initial success with CrossFire. With the Commander chemical reaction enhancement barely implemented, Beilstein announced in December 1995 that it had signed a license agreement with the Gmelin Institute to market the Gmelin database of inorganic and organometallic compounds and make it available via the CrossFire system. CrossFire Gmelin made its appearance before the August 1996 meeting of the American Chemical Society in Orlando, Florida. At the London Online meeting, Beilstein will introduce a new bibliographic database for organic chemistry to complement Beilstein CrossFire, called Netfire. Netfire, as its name implies, will be available on the Internet, but it will also be modeled as an in-house database running under CrossFire. It will be searchable by title, author, and keywords and permit display of author-produced abstracts.

While it improves and expands its databases, Beilstein is also seeking to resolve the economic issues that could yet derail its plans to reestablish Beilstein in academia. CrossFire as presently offered is arguably a good bargain, given its size and features, but even the CIC consortial rate (ca. $32,000) is well beyond the fiscal reach of all but the most generously funded universities. Building on the success of the CIC agreement, Beilstein has recently embraced a significant expansion of the academic consortial approach that evolved from discussions with the University of Wisconsin. If the Minerva project, as it is called, attracts enough subscribers (50 in addition to the CIC group), access to CrossFire*plus*Reactions will be available at Ph.D. granting institutions for $20,000. Still lower prices are under consideration for smaller colleges. The CrossFire Gmelin database will be offered under a similar agreement for $13,300, approximately half the current price for a print subscription. Initially, subscribers would access the database from a server at the University of Wisconsin, but additional mirror sites are envisioned at other universities if they are required to ensure reliable service.

Beilstein's plan to make the CrossFire databases affordable for all varieties of academic institutions is potentially very good news for anyone interested in promoting liberal access to chemical information. In the area of chemical information, the dream of many information idealists that computers would provide vastly better access to information for a broad spectrum of users has been only imperfectly realized.[23]

References and Notes

1. Reiner Luckenbach, recently retired President of the Beilstein Institute, and Clemens Jochum, Managing Director, Beilstein Information Systems GmbH, shared the Skolnik Award presented at the 210th National Meeting of the American Chemical Society, Chicago, IL, August 1995.
2. Indiana University, Michigan State University, Ohio State University, Northwestern University, Pennsylvania State University, Purdue University, University of Chicago, University of Illinois, University of Iowa, University of Michigan, University of Minnesota, and University of Wisconsin.
3. A subscription to *Tetrahedron Letters* cost $20 when it began in 1959. The 1997 price is $6845 for libraries.
4. Wiggins, G. "Using the Beilstein Database in Academic Research Libraries." Paper delivered at the 206th National Meeting of the American Chemical Society, Chicago, IL, August 26, 1993. (Abstract: *Abst. Pap. Am. Chem. Soc.* **1993**, *206*, COMP-100.) The text is available on the Internet at http://www.indiana.edu/~cheminfo/gw/beil93.html.
5. Barnard, J. M. "Substructure Searching Methods: Old and New," *J. Chem. Inf. Comput. Sci.* **1993**, *33*, pp 536–538.
6. Wiggins, G., op. cit., Abstract COMP-100.
7. Luckenbach, Reiner, Beilstein Institute, Frankfurt, Germany, personal communication, 1995; and Price, Michael, Beilstein Information Systems, Inc., Englewood, CO, personal communication, 1996.
8. Wiggins, G. "Caught in a CrossFire: Academic Libraries and Beilstein," *J. Chem. Inf. Comput. Sci.* **1996**, *36*, pp 764–769. (Abstract: *Abst. Pap. Am. Chem. Soc.* **1995**, *210*, CINF-50.)
9. Butkovich, N., Pennsylvania State University, University Park, PA, E-mail communication on CIC-BEIL listserv, February 6, 1996.
10. Barbara Allen, Director, CIC Center for Library Initiatives. CIC press release announcing the CIC/Beilstein agreement, February 6, 1995.
11. Luckenbach, Reiner, Beilstein Institute, Frankfurt, Germany, personal communication, 1995.
12. Beilstein U.S. help desk, personal communication, 1996.
13. Lawson, A. J. "Structure Storage and Retrieval using Multi-Million Files Inhouse." Paper delivered at the 206th National Meeting of the American Chemical Society, Chicago, IL, August 1993. (Abstract: *Abst. Pap. Am. Chem. Soc.* **1993**, *206*, CINF-5.)
14. CAS has offered flat-rate pricing for their entire database to some of their industrial customers, but not to any academic institutions.
15. Wiggins, G., op. cit., Abstract CINF-50.
16. See, for example *Chem. Eng. News* March 27, 1995, pp 72–76, a special issue devoted to the information revolution.
17. Consisting of one five-year index with limited search capabilities, CD-ROM products from Chemical Abstracts Service are the 12th Collective Index on CD-ROM, CD products covering certain areas of chemistry (CASurveyor), and a new, comprehensive CD product, CA on CD, which covers CA information from 1996 forward.
18. Wiggins, G.; Anthes, M. "Subsidized Searching: Summary," Wichita State University, Wichita, KA, 1996. This survey was conducted via the CHMINF-L listserv owned by Gary Wiggins at Indiana University.

19. Waldrop, M. M. "A Base Dispute," *Science* **1990**, *249*, pp 472–473.
20. Hurd, J. "Getting Started: Searching the Beilstein Database." Distributed via the CIC-BEIL listserv on September 11, 1995.
21. Twiss-Brooks, A. "Beilstein CrossFire Preparation Searching Using Commander," dated April 6, 1996. Available at http://www.lib.uchicago.edu/~atbrooks/beilstein/beilprep.html.
22. STN help desk, personal communication, 1996.
23. Subscriptions to *Chemical Abstracts* on the smaller campuses of the University of Wisconsin system disappeared in the 1970s and 1980s. There are few reports of extensive use of STN or DIALOG to access chemical databases.

Chapter 9

Use of the Beilstein System in the Chemical and Pharmaceutical Industries

Wendy Warr

A range of products from Beilstein Information Systems, including the Beilstein database, CrossFireplusReactions, CrossFire Gmelin, CrossFire Abstracts, and NetFire, are discussed with particular reference to their significance for end users in industry. Integration, end-user support, and training are significant.

Industry and Academia

The purpose of this chapter is not to duplicate material that has already been covered in earlier parts of the book, but to concentrate on issues that apply particularly to the chemical and pharmaceutical industries, and to discuss end-user searching with particular reference to the industrial environment. During 1996, there was much excitement about CrossFire in academia, for example at the Eidgenossische Technische Hochschule (ETH) (Chapter 7), the University of Indiana, the consortium centered around the University of Wisconsin-Madison in the United States (Chapter 8), and surrounding the agreement signed with the Combined Higher Education Software Team (CHEST) for U.K. universities. The uptake of CrossFire in industry was steady but received less publicity. Cynics might say that Beilstein was targeting those in universities who might be the industrial users of the future. Hard-headed vendors say privately (if not openly) that there is no money to be made from faculty and students. This author welcomes all vendors' efforts to deliver affordable services to academia and actually applauds Beilstein for recognizing the special needs of its academic users despite the prospects of financial gain being long term. Education of the next generation is the key to our future happiness and prosperity.

However, it is a fact that academics need and like to talk about their experiments, whereas industry keeps quiet until it has bought the system (and sometimes thereafter if it perceives a competitive advantage), and industrial end users of major information systems find it difficult to justify the time and effort needed to contribute chapters to books. Not surprisingly, much of the information in the earlier parts of this book has been supplied by noted academics. One of the objects of this chapter is to redress the balance. Because scientists in industry not only have little time for writing book chapters but are also unwilling to see their comments attributed to their companies in the published literature, user perceptions are often kept anonymous here. Where a company has agreed to be identified in, for example, *Beilstein Brief*, anonymity is not deemed necessary.

Inevitably, a number of German companies featured in *Beilstein Brief* have given presentations concerning their experiences with CrossFire. The German-speaking nations looked into CrossFire earlier than most, and there might be a perception in the United States that these European companies have an unusually positive attitude toward Beilstein. However, this positive attitude is not necessarily because German-speaking nations have more empathy for Beilstein for cultural and linguistic reasons. The attitudes of Swiss and German companies differ to some extent from those in the United States because the Swiss and German consortia have many years of experience of in-house access to data from the external literature. They have long recognized the importance of giving end users access to information from the literature. To put this in perspective, it should be noted that more than 50% of Beilstein's sales revenues are now coming from the United States.

Advantages of CrossFire

Many of the advantageous features emphasized in the earlier parts of this book are applicable in all working environments. Before discussing in depth the specific requirements of end users in industry, it is worth summarizing some of the unique selling points of the Beilstein Information System. One obvious one is the fact that there is no pay-as-you-go problem; users need not suffer from the so-called "taxi meter syndrome". Two great advantages of CrossFire are the interface (Beilstein Commander) and the distribution of data into the three domains, tightly coupled by hyperlinks. The domains are structures, reactions, and literature citations, each with a unique set of data fields. Hyperlinks allow seamless navigation through the domains, such as for building retrosynthetic pathways starting with a structure or reaction, and inspecting all the reactions that fit a query regardless of the document in which they were reported. Handling multistep reactions has always been a problem with reaction retrieval systems, and hyperlinks provide one apt solution. CrossFire*plus*-Reactions is a fast, inexpensive way of finding preparations. CrossFire substructure searching is remarkably fast. There are no system limits, and users can intersect huge hit lists. Both the database and the software are available for in-house use.

The Beilstein system holds Aldrich, Chemical Abstracts Service (CAS), CRC, EINECS, Fluka, Merck, and SpecInfo reference numbers, and it can be linked to ChemOffice and ISIS/Host. Beilstein is now adding ecological and toxicological data. CrossFire already has more than 350 searchable fields of chemical and physical properties. Because the factual and numeric database is key to the significance of the whole system, the next section summarizes its features.

The Beilstein Database

There is an unfortunate assumption in some quarters that the Chemical Abstracts (CA) database is so preeminent that users need no other. The CA and Beilstein databases are really complementary rather than competitive, yet there is a school of thought that regards them as alternatives. Beilstein is a tertiary information service, CAS a secondary one. Beilstein is a factual, numeric database giving direct access to data; CAS supplies bibliographies, and in the majority of cases the user has to go back to the primary literature to find data. It has been said that comparing the two products is like comparing an encyclopedia to a telephone directory.[1] Beilstein goes back to 1779; CAS goes back to only 1907 (and the Registry file back to only 1957), but it is up-to-date. CrossFire and Beilstein Online are updated quarterly and are about 6 months behind the literature.

Until 1980, Beilstein attempted to cover the entire scientific literature at the expense of currency. The policy then changed, and during the 1980s about 85 journals were abstracted. After 1990 the number increased to about 120, and from 1994, additional journals covering the fields of ecology, toxicology, and pharmacology were added in answer to user demand. The data from these extra journals will be available soon. The number of journals currently abstracted is about 180. Whereas CAS aims to cover the entire chemical literature worldwide and is not obliged to verify the quality of the material scientists publish, Beilstein has the policy of attempting to keep a high-quality factual and numeric database as up-to-date as possible. Beilstein has not covered patents since 1980, but it is the only database that covers patents before 1960. Significantly, the European Patent Office as well as the U.S. Patent Office are both users of CrossFire. The patent literature matters much more to scientists in industry than to academics, and for checking prior art it is often necessary to use multiple sources and to go back in time as far as possible. CAS covers patents and inorganics, but is also obliged to index many oddities. With the recent addition of Gmelin, CrossFire does offer access to inorganic substances.

All this raises the question of what might be missing from the Beilstein database and how reliable it is for novelty checking, that is, determining whether a compound has been reported in the scientific literature. This subject has been aired on the Internet.[2] The conclusion was that it is cheaper to look a compound up in CrossFire in-house and access CAS databases only if nothing is found in Beilstein. Recent experiments show that 90% of substances being searched for by Beilstein users are found in the Beilstein database. This is not a surprising discovery because ISI showed many years ago how few journals need to be covered to capture 90% of novel compounds.[3]

However, according to many industrial users, it is reactions rather than compounds that give CrossFire a real competitive edge. Earlier in this book (Chapter 7), Zass states that "CrossFire*plus*Reactions is not only the largest reaction database at this time, it exists also in the context of a large compound database with an unsurpassed wealth of physical data and properties." Later, he affirms that "there is no viable alternative at present regarding the time coverage and number of reactions." It would be difficult to contest this assertion. Even allowing for the fact that 5 million of the 10 million reactions are currently nongraphical, no other in-house database can begin to approach CrossFire*plus*Reactions as a definitive source of preparations. (Beilstein has announced that these nongraphical reactions will be incorporated into the graphical database by the end of 1997 as graphical half-reactions.)

In December 1996, Beilstein Information Systems announced Cross-Fire Abstracts as a key component of the citations domain. The addition of a new data field allows users to search for words contained in the title or abstract of a document. Titles and abstracts are available for all documents excerpted for the last 16 years.

CrossFire Gmelin

The *Gmelin Handbook of Inorganic and Organometallic Chemistry* not only covers inorganic compounds, it is also the largest work available on organometallic compounds. It covers crystallography, geochemistry, metallurgy, ceramics, catalysts, superconductors, solid-state chemistry, and materials science. The current eighth edition in hard copy comprises more than 700 volumes. The substance-oriented handbook is classified on the basis of chemical elements and their compounds. There are 73 volumes on iron, 62 on the actinides, 40 on boron, and so on.

The Gmelin database currently contains about 1 million compounds, including about 470,000 coordination compounds, 55,000 alloys, 14,000 glasses and ceramics, 11,000 polymers, and 3200 minerals. Factual information, including general information, physical properties, properties in/of systems, electrochemistry, and chemical properties, is contained in over 800 fields. Each fact in the Gmelin database carries a citation from either the eighth edition of the *Gmelin Handbook of Inorganic and Organometallic Chemistry* (whose entries go back to 1772), or from journals from 1984 onward. The patent literature up to 1977 is covered. Nowadays, the

112 most important scientific journals dealing with inorganic, organometallic, and physical chemistry, and with physics, are abstracted. Facts are subjected to rigorous quality testing.

The CrossFire Gmelin system has special features such as three-dimensional manipulation of structures and the Ligand Search System, allowing searching for classes of compounds, which is a valuable tool for organometallic and coordination chemists. The CrossFire structure search system had to be changed for Gmelin. For example, allowed coordination had to be increased to 20, and three-dimensional structures and stereodescriptors were also essential. Structures of compounds, substructures, facts, and citations are all searchable. Factual search includes numeric and range searching of data.

Pricing

As mentioned earlier, it is highly commendable that Beilstein believes in no second-class citizens and does not fob off academics with database subsets or "lite" software. It must be recognized, however, that some hard-headed users in industry view the academic discount as the industrial subsidy. One industrial user who remains anonymous has said: "CrossFire is an extremely useful source but unfortunately it is not cheap. Anyone coming out of university in the next few years will have become used to using CrossFire and will expect to be able to use it in any large company. They will *demand* it. So it is an essential tool, but the price problem has to be overcome. A small company could never afford it." There are, however, small companies in the United States, with as few as three chemists, who have CrossFire. In addition, small companies have the option to access and use CrossFire via the Beilstein Online database server in Colorado.

Unfortunately, CrossFire is perceived as very expensive by some companies, especially because the license fee has to be paid every year. It also seems likely that content enhancements, such as addition of toxicology and ecology data, will add a 10–20% price increase. Typical comments are: "A company that does five Beilstein searches on STN in a year cannot justify buying CrossFire," and "It's designed for companies like Glaxo Wellcome. A lot of STN Beilstein searches could be done for that money." These statements have an underlying fallacy (aside from the assumption that the pharmaceutical industry has a bottomless pit of funds!). It is very dangerous to use figures for usage of Beilstein on STN to judge how much CrossFire might be used. Zass is convinced that statistics for past usage of Beilstein on STN are not a good guide to future usage patterns for CrossFire.[4]

Beilstein Information Systems is by no means the first vendor to face the fact that one of its offerings (Beilstein on STN) is competing for revenues with another product (CrossFire) from the same company. CrossFire has features not available in Beilstein on STN, and those who state that Beilstein on STN costs less unless the Beilstein database is used a great deal are often not allowing for the cost of employing skilled intermediaries who are capable of using an online system effectively. Pricing is a really thorny issue for all database producers and their customers.

Flexibility is important. Beilstein Information Systems charges no add-on price for extra servers or CPU size. It does not accept second-class customers or task-based pricing. The highly valuable database is not separated from the CrossFire software. However, the cost of a global, multisite worldwide license, plus the need for annual payments, and the costs of support and integration are problems being addressed by some big companies. Major corporations may have much bigger budgets for information services than academia, but they are still careful about how they apportion money. The offerings of Beilstein's competitors are by no means cheap, and industry likes to see competition in the field of information products. Some companies are saving money by replacing their REACCS databases from MDL with CrossFire*plus*Reactions and by reducing the number of CAS online searches. Industry watches with pleasure as Beilstein, CAS, and MDL (and some interesting smaller players) vie with each other to produce ever more useful products, but at the same time industry bemoans the problems of integrating the competing systems!

Integration

In industry, and especially in the pharmaceutical industry, there are commonly groups of people whose main function is to provide information services, end-user support services, or information systems for research departments. The attitude to end-user searching is influenced by the skills and by the very presence of these people. Scientists may criticize high-handed systems managers who insist that the company will not support Macintoshes, or is standardizing on Microsoft solutions, and cynics may think that information professionals are merely protecting their jobs when they claim that chemists cannot be trusted to do online searching. Nevertheless, industry has many more professional information workers of all sorts than academia, and this influences attitudes to end-user searching, especially when it comes to the external literature. However, it has long been recognized that chemists need the immediacy and serendipity associated with accessing in-house information themselves, without the intervention of an intermediary. This is where systems such as ISIS from MDL Information Systems[5] have made an impact.

Multidisciplinary teams are usually involved in selecting such systems, but chemists have a strong say in any decision. Chemists' time is valuable. In the pharmaceutical industry, where there are more end users than in the chemical industry as a whole, the main mission of the chemist is getting better drugs onto the market faster, and the end user is likely to be impatient with and intolerant of slow system response, confusing menus, and systems that require extensive training. Few chemists are willing to cope with a miscellany of interfaces; many are prepared to learn how to use only one interface. "Expense is not a consideration in comparison with the importance of a single, user-friendly interface", said one computational chemist at a major European company.

CrossFire has a distinct advantage in that it can be linked to software such as ISIS and ChemOffice,[6] which are almost de facto standards in the pharmaceutical industry. Integration not only pleases the chemist who is used to drawing structures and producing reports with ChemOffice, or searching the in-house database under ISIS, but it also pleases the information staff, who do not want the overheads entailed by a proliferation of systems that need support, training, and general maintenance. The use of CrossFire with an ISIS interface does not mean that all users have to implement the same interface: any chemist who wants to use native Cross-Fire is still able to do so. Licensing the ISIS interface does incur an additional cost (in addition to licensing ISIS/Host and CrossFire). Opinions vary about any limitations (features, long-term viability, etc.) involved in taking the ISIS option.

Rüdiger Jorgensen of Novo Nordisk (*Beilstein Brief* II/96) is quoted as follows: "The strategy at Novo Nordisk is to build a comprehensive suite of databases, and the CrossFire*plus*Reactions database complements our existing investments perfectly. Running under the same environment as our other ISIS databases has made for a very fast deployment of usage, with practically no training being necessary to get the chemist started. The usage has grown enormously. . . . CrossFire*plus*Reactions has been received very positively."

Constantin Zirz (*Beilstein Brief* 1/95) lists the ability of the Beilstein CrossFire system to be integrated, its up-to-dateness, and its extensive data coverage as the most important reasons for its use at Bayer. Zielesny et al.[7] have described Bayer's Integrated Chemistry Information System, which uses the ISIS graphical user interface for access to both proprietary in-house information and the Beilstein database. An end-user crossover application has also been developed where a structure, or structures, from an ISIS hit set can be used to search online databases on STN such as CAPlus, ChemInform, and CASREACT. Donner[8] has reported on the economic aspects of the Bayer system. He also makes a plea for wider use of CAS Registry Numbers (of which there are many in the Beilstein database) in the interests of the worldwide chemical community.

While not commenting on the value of Registry Numbers as CAS proprietary information, R. L. Swann of CAS has described the integration of public and proprietary information as "the Holy Grail for the chemist".[9]

The CrossFire option for accessing both internal and external databases has already been adopted by many companies, and CAS has now entered the arena with the Chemical Abstracts Gemini project currently being developed for Kodak. MDL is also known to be interested in the topic (in a wider sense than the current close integration of ISIS and CrossFire). The significance of World Wide Web browsers in integration is discussed later in this chapter.

A recent research project suggests another possibility for integration in the pharmaceutical industry. One of the aims of de novo drug design software is to suggest new compounds for the chemist to synthesize, that is, compounds that are significantly different from those in the company's compound collection. Combinatorial explosion is a problem with structure generators in such software, and one way of reducing the enormous number of drug candidates is to evaluate which ones are synthetically feasible. Algorithms calculating scoring functions for synthetic feasibility have been devised.

Carino et al. at the University of Wisconsin-Madison, in conjunction with Beilstein Information Systems, Inc., have tackled the problem of synthetic accessibility by automating the process of searching the literature for synthetic procedures.[10] They generated 20,000 compounds using GrowMol, to fill the P1, P2, and P3 pockets of the active site of porcine pepsin. After energy minimization in the active site and rejection of structures with high conformational strain energy, they examined 2400 unique structures for complementary contacts with the enzyme. They selected the best 300 compounds and performed automated searches of synthesis data in the Beilstein database using CrossFire*plus*Reactions. The structures for which no literature citation was found were then modified, and a new, more general search for their analogues was carried out. Automation of the literature search process using molecular structures generated by computer programs gives the user access to more diverse synthetic procedures, for a wider variety of compounds, in a shorter amount of time.

Supporting End Users

End users in industry (and their information systems and services staff) have very high expectations in terms of user friendliness. They expect especially intuitive interfaces and lightning response times. The companies that have installed CrossFire seem well satisfied with the user interface, as exemplified by comments in this chapter, and the substructure searching is certainly incredibly fast. However, there have been some suggestions for improvements in the Fact Editor and for a reduction in the number of menus. Range searching of numeric data can be slow. Data export may be a problem: clumsy clipboard export or the "print to file" option to make a PostScript file of a diagram plus facts do not meet the needs of industrial users for putting structures in spreadsheets.

Most large companies have staff dedicated to supporting end-user systems, that is, people who install software, update databases, run training courses, staff help desks, and so on. They are also involved in decisions on software and hardware. In some CrossFire trials, chemists have been deterred by slow Internet links to the Beilstein server. It is up to the support staff to ensure that a proper trial takes place and to specify hardware, software, and network needs.

In the early days of CrossFire, companies experienced some hardware problems. There was no Macintosh client at first, which caused more dismay in the United States than in Europe. The first Macintosh implementation also had some teething troubles, but these have been addressed. At first an IBM RS/6000 server was obligatory. Some companies would not accept an RS/6000 in their infrastructure, however, because there were few other research systems that they could run on it. Nowadays, CrossFire will run on DEC Alpha under Open VMS and on a Windows NT server. A decision as to whether to support Silicon Graphics machines may be made in 1997.

Support staff are also responsible for installing new versions of software and databases. Frequent upgrades and changes can be time consuming and may deny users access to a mission-critical system. Very frequent changes are usually not acceptable. At Boehringer Mannheim, updates are done quarterly by exchange of the disk cabinet (see *Beilstein Brief* II/96). This means that users suffer only 15 minutes of down time. For reasons of time and security, the transfer of data by magnetic tape was rejected. End-user support staff are also responsible for training and for running help desk services. The more intuitive the system, and the better the training, the less pressure there will be on help desk staff.

Training

Tine Buchkremer of Byk Gulden has stated (*Beilstein Brief* I/96) that "CrossFire . . . requires almost no training for the end user. . . . No one should be satisfied today with a new system for the retrieval of chemical information which requires extensive training for the end user." As mentioned earlier, Novo Nordisk reduced the need for training by adopting the familiar ISIS interface. In my opinion, it is essential to invest in good training for CrossFire, but the amount of training needed will vary depending on the needs and experience of the user. There are two reasons behind this assertion. First, just a small amount of training will avoid some real pitfalls (examples follow), and yet more training will repay itself in terms of faster or higher quality retrieval. Second, it is a truism that the more powerful the features offered by a system, the more complex it will be to use (assuming that the chemist wants to take full advantage of most of the facilities). The CrossFire interface is intuitive in many respects, but this does not mean that training can be avoided altogether.

To do structure entry, and substructure and reaction search, the user needs some grasp of the conventions used in the system. Chemists already familiar with some other major software systems will need to be told that CrossFire uses the "free site" approach. Novices will need to be taught the difference between reactions, which are the chemical behavior of a reactant, and preparations, which are ways of making a product. Users in the United States, particularly, have to beware of using American spelling. It is best to assume that both British and American spellings (and misspellings) may occur and that chemical names may appear without the expected "e" on the end, for example, because Beilstein is likely to reflect exactly what the original author published.

As a real life example of pitfalls awaiting the novice, I (having a particular interest in combinatorial chemistry), with some assistance, decided to try a search for the well-known benzodiazepine syntheses carried out (independently) by Jonathan Ellman and Sheila DeWitt. Unfortunately, the name "benzodiazepin(e)" is ambiguous, even though it is associated with one specific ring system in the pharmaceutical industry. By selecting the name "benzodiazepine" from the thesaurus, we accidentally pasted the wrong structure into our reaction search. Even when we had corrected this error and tried searching for the correct structure as a reaction product, together with the authors' names, we still did not find the required reactions. We ultimately concluded that there were two likely reasons for our failure. First, Ellman published his products as a generic or "library" structure, and his reactions are thus in CrossFire*plus*Reactions as nongraphical reactions. Second, DeWitt's work was published in *Proc. Natl. Acad. Sci. USA*, which is not abstracted by Beilstein because it is not a mainstream organic chemistry journal.

Another example is a search for compounds with antifungal or antimycotic effects, such as terbinafine, which belong to the allylamine category. There are quick ways of entering the terbinafine structures, and there are inefficient ways. Stereochemistry then has to be specified at the double bond. A stereobond is represented with an "St" on it in the structure display. In setting up a stereosearch, it is necessary, for example, to understand the definitions of "absolute", "relative", and "racemic". The terbinafine search could be continued by combining the structure query with a data query:

$$bf = antimycot^* \text{ or } bf = antifung^*.$$

where bf is the abbreviation for the required data field, and the asterisk is a truncation symbol. Finding the required data field is easy, even though there are 350 of them, once a user has been shown how to do it. The search could also be carried out by running the structure query first to give a hit set (q02), and then combining the hit set with the factual query:

[bf = antimycot* or bf = antifung*] and .q02.

The order of factual queries can be important. Consider a structure query that has given a hit set q08 with too many answers, and the user wishes to look at only the 1990 literature. A search

.q08 and py = 1990

defaults to the reaction database, whereas a search

py = 1990 and .q08

defaults to the citation database.

The importance of atom–atom mapping in reaction searching has been described elsewhere in this book. It is another of the features that chemists need to learn to get the best out of the system. Users are going to be more efficient searchers if they are properly taught all the above concepts.

NetFire

The incredible impact of the Internet and the World Wide Web, and the implementation of intranets in many companies, means that a Web browser interface now holds much appeal for end users.[11] Beilstein Information Systems recently launched its first product to take advantage of the Internet, a new literature-based database and search system on the Web called NetFire,[12] access to which will be free of charge until the middle of 1997. Thereafter, NetFire will be offered by annual subscription. The system is designed to take advantage of features found in the most popular Web browsers, and it gives access to titles, abstracts, and authors of articles published since 1980 in the same journals that are abstracted for CrossFire. The query-by-form mechanism permits the formulation of a great variety of queries. Users may search by author, or by words included in the abstract or title. The search may be restricted to a certain journal or time range. A table of results is produced, listing authors, title, literature references, and the abstract, with the queries highlighted in a contrasting color (so-called hit term highlighting).

My first and very pleasing experience with NetFire was the simplicity of finding (at no expense) all the papers by Jonathan Ellman that had proved elusive in the ill-constructed CrossFire search outlined in the previous section. It will be interesting to see how Beilstein Information Systems builds upon products such as NetFire (for example, by adding NetFire–CrossFire links) as it continues to develop its system for the 21st century.

References

1. House, H. O. *J. Chem. Inf. Comput. Sci.* **1984**, *24*, 277.
2. "Beilstein vs. Registry" is accessible via WWW at http://www.mailbase.ac.uk/lists/chest-crossfire/1996-07/0003.html and the following pages.
3. Garfield, G. "Index Chemicus Goes Online with Graphic Access to Three Million New Organic Compounds," *Current Contents* June 25, 1984.
4. Zass, E. Re: Beilstein CrossFire accessible via WWW at URL http://atlas.chemistry.uakron.edu/CHMINF/chminf_jul96/0139.html.
5. ISIS is the Integrated Scientific Information System from MDL Information Systems, Inc., 14600 Catalina Street, San Leandro, CA 94577, USA, tel. 510-895-1313, fax 510-352-2870, http://www.mdli.com.
6. ChemOffice is a desktop publishing and chemical structure drawing system from CambridgeSoft Corporation, 875 Massachusetts Avenue, Cambridge, MA 02139, USA, tel. 617-491-2200, fax 617-491-0555, http://www.camsci.com.
7. Zielesny, A.; Becker, M.; Köhler, A.; Sendelbach, J.; Zirz, C.; Donner, W. T. "Scientific PC-Based End-User Information Workspace: New Developments at Bayer." In *Proc. 1996 Int. Chem. Inform. Conf.*; Collier, H., Ed.; Infonortics: Tetbury, Gloucestershire, U.K., 1996; pp 125–123.
8. Donner, W. T. "Economic Aspects of Chemical Information," *J. Chem. Inf. Comput. Sci.* **1996**, *36*, 937–941.
9. Swann, R. L. "CAS Information Systems: Managing Technology Today, Creating Technology for the 21st Century." In *Proc. 1996 Int. Chem. Inform. Conf.*; Collier, H., Ed.; Infonortics: Tetbury, Gloucestershire, U.K., 1996; pp 131–134.
10. Carino, S. E.; Cieplak, T. P.; Dales, N. A.; Rich, D. H.; Manrique, J. "Automated Literature Searching of the Beilstein Database: The Synthesis of Molecular Structures Generated by GrowMol using CrossFire*plus*Reactions." Poster presented at the ACS National Chemical Information Symposium at the College of Charleston, Charleston, SC, July 14–18, 1996.
11. Jochum, C. "The Web and Chemical Information: New Prices, New Features, New Knowledge?" In *Proc. 1996 Int. Chem. Inform. Conf.*; Collier, H., Ed.; Infonortics: Tetbury, Gloucestershire, U.K., 1996; pp 121–123.
12. Heller, S. R. "The Beilstein NetFire System," accessible via WWW at URL http://www.elsevier.nl:80/section/chemical/trac/netfire.htm.

Chapter 10

AutoNom

Janusz L. Wisniewski

A person with one watch knows what time it is;
a person with two watches is never sure.

Proverb

This chapter describes AutoNom, a fully automatic and practical computerized system for the generation of IUPAC (International Union of Pure and Applied Chemistry) systematic nomenclature directly from graphic structure input. The algorithm developed for AutoNom analyzes the compound's structural diagram, input via a graphic interface, and generates the name purely on the basis of the resulting molecular connection table. This chapter discusses the fundamental aspects of AutoNom's naming algorithms and the programming approach used, and gives an overview of the performance and accuracy of the computer system.

The need to precisely describe information on the structure of chemical compounds spans written and oral communication in a multidisciplinary, international environment. A variety of methods have been devised to impart information on the structural representation of compounds. These methods include the use of molecular formula (MF), various line notations, trade and trivial names, registry numbers (RNs), structure diagrams, and systematic nomenclature.

Molecular formulas define the atomic composition of compounds, but not the relative positions of the atoms. In addition, molecular formulas are not specific to a single compound, but may be identical for many compounds with different structures.

Various line notations have been devised to accurately depict both the atomic content and orientation of compounds. Much of the focus in creating notation systems has been geared toward computer input and manipulation of structures. Some of the well-known notations include Wiswesser (WLN), Dyson–IUPAC (International Union of Pure and Applied Chemistry), Hayward, Skolnik, Gremas,[1] and, more recently, SMILES[2] and ROSDAL.[3] All of these systems require detailed memorization of their rules and conventions and are limited strictly to the audience familiar with those rules.

© 1998 American Chemical Society

Trade names and trivial names are one of the oldest ways of communicating information on compounds.[4] The use of trade names, however, suffers in international communication and in accurately describing the compound. A name such as "Red Dye No. 8" does little to describe the structure or type of compound being discussed.

Over 15 million organic compounds have been described in the course of the past 155 years, and the number increases from day to day. In the early 1960s, Chemical Abstracts Service (CAS) started to give each compound in its computer system database an individual number called the CAS Registry Number. Registry Numbers are a convenient identification system for tracking information and providing a key to look up data, but they impart absolutely no structural details in moments of communication. Besides, they are fully linked with the database systems for which they have been generated and convey only information on the relative position of the structure in the specific collection of structures, and nothing more.

For the practicing chemist, the most acceptable visual representation of a chemical compound is a two-dimensional plan of the three-dimensional structure, and this usually gives an adequate view of its structure, easily understood, easily drawn and copied. Used in a reaction sequence, it shows clearly the course of the reaction, with display of transient intermediates if required. If it is necessary to show the third dimension, this can be done by using special bond configurations representing lines going above or below the plane. This diagrammatic representation is usually referred to as a structure diagram or structure graph. Structure diagram conventions are established as an international standard, and they are a unique method of communicating information on a chemical compound. For computer processing, the diagrams are transformed (manually or automatically) into linear strings of characters or into two-dimensional matrices listing all the atoms (nodes) and their mutual interconnections. These concise representations are referred to as connection tables (CTs). Although important only for computer storage and processing, CTs have recently become the main means of communicating information on chemical compounds.[5] Uniquely derived from the standardized structure diagrams, CTs have become the most complete structural representation and, at the same time, the most visually unrecognizable to humans and the most computer friendly.

The principal use of chemical nomenclature is to give a compound a label that can be spoken, written, and used in printed indices and from which the structure can be perceived by scientists. Although trivial nomenclature has the benefit of conciseness, only systematic nomenclature, which to a certain extent gives pronunciation and semantics to a structure, is of use for unambiguously labeling a structure with a name that can safely be communicated worldwide. It is the aim of systematic chemical nomenclature to describe the composition, and insofar as practicable, the structure, of compounds.

Hanging over the great complexities and illogicalities of current nomenclature is the shadow of the computer. The marshaling of almost 7 million structures (the Beilstein structure database reached this size in 1997) and their attendant properties is increasingly the responsibility of computer systems. Because computers depend on logic, their use promotes systematic nomenclature. The nomenclature that is regarded as "systematic" today is defined by the consensus of users' opinions. As in all linguistics, there is a struggle between pragmatists, who regard as satisfactory any word that conveys the intended meaning, and purists, who insist that rules ought to be followed, with, unfortunately for the computer programmer, the pragmatist having the advantage. Thus, the Commission on the Nomenclature of Organic Chemistry (CNOC) of IUPAC, which is responsible for inventing, monitoring, and revising the recommendations that are guidelines to systematic nomenclature, tries to see nomenclature as a whole, codifying already existing usage into rules and only very occasionally suggesting novelties. The IUPAC system has been developed for over 75 years and is far from perfect. What is even worse, it has never become a universal standard. The list of reasons is long, but the top one is ambiguity. In times of computer usage, this is the major obstacle to standardized, systematic nomenclature. And the need for such a nomenclature exists.[6]

To address the problem of ambiguity in the selection of preferred names, the two most important producers and distributors of chemical information, CAS and the Beilstein Institute (Beilstein), devised undocumented, ad hoc subrules that only amplified the difficulty of uniquely naming organic compounds. These rules were necessary because IUPAC recommendations frequently allow more than one name for a given chemical compound. As a result, both institutions revised the IUPAC system and created their own "systematic" IUPAC-compatible nomenclature rather than IUPAC-sanctioned nomenclatures. In addition, trivial or trade names, being shorter and more concise, have successfully replaced systematic names for several chemical compounds that are of commercial importance or are the subject of public concern, such as pharmaceuticals, insecticides, and pollutants.

Both CAS and Beilstein claim to conform to the IUPAC rules, and in general they do. The IUPAC recommendations were consciously formulated to allow considerable freedom in their application, and in many cases they are not fully defined to their logical conclusion. In practice, this means that any given structure does not necessarily relate to one unique correct name. Thus, the specific "dialects" supported separately by CAS and Beilstein can still represent systematic nomenclature no matter how far apart they are. This, as far as computer usage is concerned, is the greatest weakness of the language. Taking all these facts into account, a strong reluctance by the chemical community, and in particular the chemical industry, to perceive the need for fully systematic nomenclature is understandable.[7]

The "dialects" of IUPAC are nowhere clearly defined for an average user. This has also contributed to the difficulties of establishing an unambiguous nomenclature standard. As long as such a standard does not exist, practicing chemists will find themselves alienated to a great extent from systematic nomenclature. Theoretically, even if some sort of consensus were achieved and the unambiguous nomenclature standard worked out and adopted, there would still be a problem of nomenclature complexity. It is generally accepted that IUPAC nomenclature is cumbersome, with a very large number of rules that are often difficult to follow. Frequent alternatives allowed in name assignment, contradictory recommendations, the lack of rules in certain areas, and the exaggerated freedom in interpretation of the rules lead to ambiguity and specific nomenclature chaos. It has recently become well known in organic nomenclature circles that a co-worker of the Beilstein Institute (Frankfurt/Main, Germany) has produced 29 systematic, fully "correct" names, as far as IUPAC recommendations are concerned for a single structural diagram. Sarcastic organic chemists claim that synthesizing a new organic compound is sometimes easier than assigning a name to it.

The basic problem of naming is that a correct name is not necessarily the *only* correct name for a structure. To complicate matters, the rules for arriving at a correct name, as was already pointed out, are complex, and very few chemists can handle them. Even worse, the most important centers for chemical documentation in the world are not uniform, either internally or externally, in their treatment of the rules. This is not the result of carelessness or lack of effort, it is simply a reflection of the difficulty of agreeing how a multidimensional problem can be forced into a single, universal text description. For example, Figure 10.1 shows the structure of a compound and its many names.

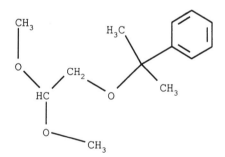

1,1-dimethoxy-2-[1-methyl-1-phenyl-ethoxy]-ethane or
[1-methyl-1-phenyl-ethoxy]-acetaldehyde dimethyl acetal or
((,(-dimethylbenzyloxy)-acetaldehyde dimethyl acetal or
[1-(2,2-dimethoxyethoxy)-1-methylethyl]benzene (CAS name)

Figure 10.1. A typical organic compound and its many names.

In principle, there is nothing wrong with a multiplicity of names for structures. As long as each name is an adequate representation of the structure, there are no real problems, apart from ensuring that chemists are reasonably familiar with the rules in a passive sense (i.e., they can interpret a name as opposed to creating one). However, the traditional attempted use of nomenclature has been much greater in its scope. Before computerization, the ideal was to index every significant structural sub-unit of the structure using nomenclature. The structure would be intu-itively broken down into areas of relevance (e.g., acetaldehyde, benzene, ethane), and these were bound together into a text by use of locational parameters (1, 2, α, and so forth). This approach is based on chemical experience and is by no means bad. But it contains the limits of its own applicability in so far as the vocabulary used has never been fully stan-dardized in a strictly defined sense and the intuitive subdivision has never been fully cleared of internal contradictions. This has meant that the use of indices based on names or parts of names remains to this day a haz-ardous business. To use the above example, it is not immediately obvious to most chemists whether they should be looking under A (for acetalde-hyde), B (for benzene), or E (for ethane). A computer system capable of generating names algorithmically, and consequently using the same rules of relevance, would always lead to the same index name, thus solving the problem once and for all.

The absolute contradiction between the uniqueness of structural dia-grams, their respective connection table equivalents, and the complexity of naming organic compounds led the author of this chapter to investigate the feasibility of their mutual interconversion. Particularly interesting was the direct conversion of structural diagrams into systematic chemical names.

Since the pioneering work of Garfield[8,9] in 1960, it was always a challenge to create an automated computerized system that would be able to implement the direct conversion of names to formulas and formulas to names. For a long time, unfortunately, it remained a challenge. In 1981, CAS started a project that was supposed to create the first computerized system for automatic translation of chemical structures into chemical sys-tematic names[10] (using a CAS-based system, obviously). Besides the chal-lenging scientific aspect behind the project, there was an important economic one. CAS, and similarly Beilstein, has a daily influx of 1000–1500 new compounds to be named.

The predictions made by the authors involved in the CAS project were anything but encouraging. After six person-years of design-phase work, they concluded, "the programming algorithm will likewise require a siz-able effort, estimated as 8–10 work years" (reference 10, page 192). The project was never completed, and there are no reports by CAS about successful implementation of the complete system.

In the beginning of 1987, the Research and Development Department at the Beilstein Institute (Frankfurt/Main, Germany) launched its own project aimed at providing the chemical community with a fully computerized "structure-to-name" translator. The system, when completed, would take the task of nomenclature back to the basic principles, and thus would relieve the chemist of the difficult choices involved in the application of the nomenclature rules. Besides the commercial interest in producing a tool for the Nomenclature Department at Beilstein, it became obvious from the beginning that the most important advantage of the computerized nomenclature system would be standardization of naming principles. That would doubtless be the major positive result once the use of the system became widespread.

Four years later, the first version of the system was complete. It was immediately integrated into the production scheme of the *Beilstein Handbook* and the Beilstein database. The Beilstein database was extended, and the names generated by the system become a normal searchable data field in the Beilstein file offered by both the DIALOG and STN information networks. Later, they also became fully searchable in the Current Facts CD-ROM service as well as in the largest in-house chemical database system, CrossFire, which was offered by Beilstein starting in 1994.

A few months later, it was decided that the usefulness of the system was appreciated by the chemical community, and after a technical remake in February 1991, the system was offered as a commercial software package under the name AutoNom (from *Auto*matic *Nom*enclature).

The system's standardization potential as far as systematic nomenclature is concerned was immediately noticed. In 1991, the IUPAC CNOC, during its meeting within the General Assembly of IUPAC in Hamburg, Germany, invited the author to introduce the system in front of the Commission. During this meeting, a new major project of great complexity, called Preferred Name (or P-Name), was announced by IUPAC. Its purpose is to provide, for the first time in IUPAC's long history, a rigorous protocol to generate *a single approved name for each substance*. In special situations such as regulation of trade, patent law, and emergency hazard management, unique names are necessary to permit nonchemists to deal unambiguously with chemical substances. It was confirmed by the Commission that AutoNom is implementing most of the relevant recommendations of such a rigorous protocol, and the Commission decided to closely monitor and support further development of the system.

After its introduction in 1991, the system soon started to become a useful tool for the chemical community. In 1992, four independent and most favorable reviews of the system appeared in scientific journals in the United States[11,12] and Europe[13,14]. All four publications recommended the system as a useful tool not only for producers of chemical information, but also for research chemists and university teachers of organic chemical nomenclature.

One year later, the first careful and detailed evaluation of the system was conducted in a big industrial chemical company. Wyeth Ayerst Research (Princeton, NJ) evaluated the system for use in naming more than 10,000 organic compounds from its internal corporate database, and it presented the results[15] at the 206th National Meeting of the American Chemical Society, in Chicago, IL, August 22–27, 1993. In a very favorable report, the first independent statistical results of AutoNom's success rate were presented. For the 10,000 compounds at Wyeth Ayerst, AutoNom could name up to 90% of them, which is more than the average success rate of the system (ca. 85%) for a random variety of organic compounds.

Recently, the Sigma-Aldrich Chemical Company (Milwaukee, WI) licensed the AutoNom program and plans to publish,[16] in its famous catalog, the names generated by AutoNom in addition to the current names.

After the system was introduced, the scientific progress that resulted was reported in a number of publications[17-23] and presented at a number of chemical meetings, congresses, and conventions. In addition, the system proved to be a commercial success. The competitive market of software providers in the area of chemical information is the best test of system capabilities, usefulness, and acceptability. It was purchased in the meantime by several hundred chemical institutions, including both university users and major chemical and pharmaceutical giants, such as Ciba-Geigy, Merck, Hoechst AG, BASF, Bayer AG, Glaxo Laboratories, Pfizer, Boots, and L'Oréal, to name only a very few from the top of the list.

Systematic Chemical Nomenclature

In contrast to the IUPAC nomenclature of organic compounds,[24] other systematic nomenclatures (such as Lozac'h nodal nomenclature[25] and Hirayama's HIRN system,[26] to mention the two best known) proposed in the course of the last 30 years were never contenders for international or even local acceptance. Even if they currently have no practical meaning, it is interesting to outline their existence here.

Systematic nomenclature is a set of rules or recommendations that define, with examples, the permitted forms of construction and the accepted trivial or formal terms from which names may be formed. The rules, when followed, should give a complete, unique, unambiguous representation of the structure of an organic compound. Of the systematic nomenclatures, only one is recognized internationally. This is the IUPAC nomenclature for organic chemistry, which is the definitive nomenclature used in most scientific works. The nomenclature is based on so-called recommendations, also known as rules, which are under constant review by CNOC, an international body of nomenclature chemists invited by IUPAC. New proposals are published in the journal *Pure and Applied Chemistry*. Because it is an international nomenclature, it has variants in languages other than English, but the rules are universal for all natural languages.

The first monograph (in English)[27] on systematic nomenclature appeared in 1958. It covered codified recommendations for hydrocarbons (Section A) and heterocycles (Section B). During the 1960s and 1970s, CNOC continued to publish new "sections" of its nomenclature compendium and revise older ones. In 1979, the latest versions were combined in a monograph[24] that is commonly called the *Blue Book* and which is presently the only official handbook of nomenclature of organic chemistry.* The most important recent monograph edited by CNOC is *A Guide to IUPAC Nomenclature of Organic Compounds, Recommendations 1993,*[28] a comprehensive guide to organic nomenclature. The *Guide*, which makes significant changes and extensions in several areas, is mainly intended as a complement to the *Blue Book*, not a replacement for it.

Computerization of Organic Chemical Nomenclature

The development of computer methods for the interconversion of chemical nomenclature to and from molecular formulas, connection tables, and structural diagrams has followed and continues to follow two separate paths. On the one hand, there are a great many reports, mainly from university sources, dealing with translation of systematic names into structural diagrams. On the other hand, there is relatively limited literature on the translation of structural diagrams directly into systematic chemical names. Although these are reverse directions of the same conversion, they have in practice very little in common as far as algorithms and applicable methods are concerned. The only problem they share, even if it sounds sarcastic, is the complexity and ambiguity of the systematic nomenclature.

In the case of a name-to-structure translation, the unambiguous and to some extent undefined entity is the input, i.e., a name with all its possible "dialects", allowed notations, and lack of sharpness of syntax. The output—the structural diagram—is, on the other hand, unique and is defined to the smallest detail. Any practical general-use system for translation of chemical names into structure, and such a system does not yet exist, could only function with a strict and distinct specification of the allowed and approved syntax of the input name string. This would limit the use of such a system to only a particular type of systematic nomenclature, putting in doubt the feasibility of a commercial system. The first

---

* The numerous publications that contain word "IUPAC" in their titles, and treat chemical nomenclature, no matter how often they refer to the *Blue Book*, are under *no* circumstances meant to replace the *Blue Book* itself. The updated version of the *Blue Book* is in preparation by the Commission.

functioning system of this type, called VICA, was developed by Domokos and Goebels[29] for the mainframe computer in the Beilstein Institute. It has been successfully applied in Beilstein (reaching a success rate of up to 95%) for Beilstein nomenclature, and has never been used outside Beilstein.

A discussion of the details of the conversion of chemical names into structures is beyond the scope of this chapter. The interested reader may consult the excellent series of expert publications by the team at the University of Hull.[30–33]

It must be stressed here that other than AutoNom, there is no general computer-assisted nomenclature system for automatic conversion of structural diagrams of organic compounds into chemical names. The reports on research in this area present systems, methods, or simply theoretical studies that are strictly limited to certain classes of compounds (e.g., only selected types of ring systems or only alkanes with no hetero atoms). Therefore, the issue of computer-based comprehensive and global translation of structures into nomenclature remains open.

The earliest relevant paper by Conrow[34] on research in this area dates back to 1966. He designed a Fortran program for the computer generation of IUPAC names for bridged bicyclic, tricyclic, and tetracyclic hydrocarbons. Conrow's program was the first effort to translate connection tables directly into IUPAC nomenclature. The program was described by the author as the implementation of nomenclature for "von Baeyer hydrocarbons". Its most severe limitation was that polycyclic compounds having more than four rings were excluded. Furthermore, in the case of a tetracyclic compound that is topologically twisted, absurd names were generated, mainly because of severe problems with treatment of highly symmetrical systems with many alternative bridges. In addition, quaternary atoms in the molecule under consideration caused the program to cease its operation.

In noting the deficiencies of Conrow's program, van Binnedyk and Mackay published a paper[35] seven years later on a much more comprehensive program for automatic name generation of von Baeyer nomenclature for hydrocarbon ring systems. The algorithm was not limited by the number of rings. The only limitation was the computer technology of the time.

The problem of computer-assisted von Baeyer nomenclature has, after almost 20 years, been revisited again by Rücker et al.[36] They describe a computer program that generates IUPAC von Baeyer names and the corresponding numbering schemes for polycyclic hydrocarbons of any size and complexity. The program POLYCYC, developed in Freiburg University (Freiburg, Germany), is the most efficient general-use program of this type in terms of computer resources.

It is interesting to note here that Babic et al.[37] have recently developed an algorithm for systematic IUPAC naming (using the von Baeyer convention) of spherical carbon molecules called fullerenes.* They studied fullerene isomers with up to 70 carbon atoms by using what they dubbed Hamiltonian cycles,[38] that is, cycles characterized by passing through each atom in the structure exactly once. This property facilitates the finding of IUPAC names because one knows in advance the size of the maximal ring, which appropriately narrows the search. It is worth quoting the von Baeyer name obtained for so-called *buckminsterfullerene*.[39] For this C_{60} fullerene with 33 double bonds distributed in 32 subrings, the following name, gibberish for a normal chemist, as cited in reference 37 is:

Hentriacontacyclo[29.29.0.$0^{2,14}$.$0^{3,12}$.$0^{4,59}$.$0^{5,10}$.$0^{6,58}$.$0^{7,55}$.$0^{8,53}$.$0^{9,21}$.$0^{11,20}$.$0^{13,18}$.$0^{15,30}$.$0^{16,18}$.$0^{17,25}$.$0^{19,24}$.$0^{22,52}$.$0^{23,50}$.$0^{26,49}$.$0^{27,47}$.$0^{29,45}$.$0^{32,44}$.$0^{33,60}$.$0^{34,57}$.$0^{35,43}$.$0^{36,56}$.$0^{37,41}$.0^{38} ^{,54}.$0^{39,51}$.$0^{40,28}$.$0^{42,46}$]hexaconta1,3,5(10),6,8,11,13(18),14,16,19,21,23,25,27,29(45), 30,32(44),33,35(43),36,38(54),39(51),40(28),41,46,49,52,55,57,59-triacontaene.

The authors[37] admit that for the gigantic and famous C_{380} fullerene, which also possesses Hamiltonian cycles, the IUPAC name could not be derived owing to computer limitations.

A similar program with extension for heterocyclic systems with replacement nomenclature (Recommendation B-4) was reported in 1990 by Röse.[40] The additional module, which was able to handle replacement nomenclature for polycyclics containing hetero atoms in rings, was an important advance over the program by Rücker et al. The program, known under the name BAEYER, accepted input structures in MOLFILE format and was interfaced with its own structure editor, which enabled input of the ring systems to be named. The output of the system consisted of the name as well as the matrix of IUPAC-recommended locants for all input atoms.

Automatic computer identification of alkanes was reported in 1989 by Davidson.[41] His Fortran program has implemented, although with many restrictions and limitations, Recommendations A-1 through A-4 for nomenclature of acyclic hydrocarbons.[24] The program identified all the alkane chains in a structure and was able to select a master chain, according to 1957 rules only, and order the other chains as substituents. Any introduction of functionalities containing hetero atoms or any, even the simplest cyclic characteristics, led to rejection by the algorithm.

The feasibility of creating a microcomputer-based program that would accept structure input and generate an IUPAC-like systematic name was extensively researched by Meyer and Gould.[42] In order to prove their thesis they designed and wrote a simple BASIC program that identifies graphically entered structures. The program works strictly on a lookup basis, matching fragments of structures (parent structure and

---

* Although not strictly defined, fullerenes are thought of as spherical molecules with only five- and six-membered rings. The best known, and recently most researched, fullerenes start at C_{60} and end at the "monstrous" C_{380} sizes.

localized substituents) against the fragments prestored in the name fragment dictionary. The dictionary was limited initially to only 860 structure/name fragments selected from the most commonly utilized substructure fragments as identified in the *ISI Chemical Structure Dictionary*.[43] Because of the very limited dictionary, the program could be used only for strictly prescreened types of structures and was more of experimental significance than practical use.

Interesting educational software for helping students of organic chemistry at Eastern Washington University (Cheney, WA) learn to use IUPAC rules was reported by Raymond.[44] The program, written in LISP, proved to be effective in naming alkanes, alkenes, alkynes, and their corresponding halides. Monocyclic molecules ranging in size from three- to six-membered rings are allowed. For acyclic compounds, moderately branched hydrocarbon chains up to 10 carbon atoms in length are accepted. No functional groups containing hetero atoms are allowed.

Another experiment of this type is the BEAKER program,[45] developed recently at Northwestern University (Evanston, IL). This bundled software package,* meant as a tool for chemistry students, offers as a utility a nomenclature generator, which surprisingly works in both directions: structures directly from IUPAC names and names directly from structures. The system's abilities, however, are highly limited: names are generated only for extremely simple molecules. Rings are limited to a very few monocyclic alkanes (hetero atoms in rings are not allowed), the number of recognized functional groups does not exceed 50, and even for chains the program rejects naming compounds by heavily substituted chains.

An interesting study more of theoretical than of practical use, on linking a nodal notation of a structure with systematic nomenclature, was recently presented by Navech and Despax.[46] The complex algorithms expect structures to be first coded into so-called SCN code (Structure Code Number). The SCN generated for the whole molecule is then processed, and nomenclature is generated. The capabilities of the package are limited to simple acyclic and monocyclic compounds, with no hetero atoms and multiple bonds inside chains allowed.

Automatic Name Construction from Structural Diagrams

The algorithm designed for the AutoNom program analyzes the structural diagram of an organic compound entered via a graphic package and generates a chemical name purely on the basis of the connection table derived from the input structure. The system is delivered as a software package together with the bundled structure editor, which enables drawing or editing the structure to be named and submitting it to the naming

---

* The program (for Macintosh computers) is commercially available from Brooks/Cole Publishing Company (Pacific Grove, CA).

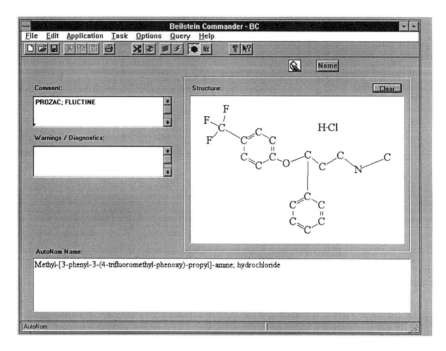

Figure 10.2. User interface of AutoNom program.

routine. The format of the connection table derived for a graphical representation on the screen depends on the structure editor used. The program is so designed that the interface with the graphic module is fully separated from the operational module responsible for naming. The connection table leaving the structure editor and delivered as input to the algorithm is transformed into internal AutoNom specific connection table representation, especially designed and optimized for naming. The version offered currently is capable of accepting structures represented, as connection tables, coming from the two most important and widespread structure editors, namely, Beilstein structure editor in native ROSDAL format[3] as well as in MOLFILE[47] format, which is known from the popular drawing package ISIS/Draw from Molecular Design Limited (MDL). The capture of the user interface screen (under Beilstein Commander) program for the currently available Windows version 2.0 of AutoNom is presented in Figure 10.2.

During the setup phase, the structure connection table (delivered by the currently active structure editor) is handled by the input interface and coded into a connectivity matrix (CM), which specifies which atom is connected with which other atom and establishes the type of bonding, atomic number vector (AN_v), and charge vector (CH_v). These three data

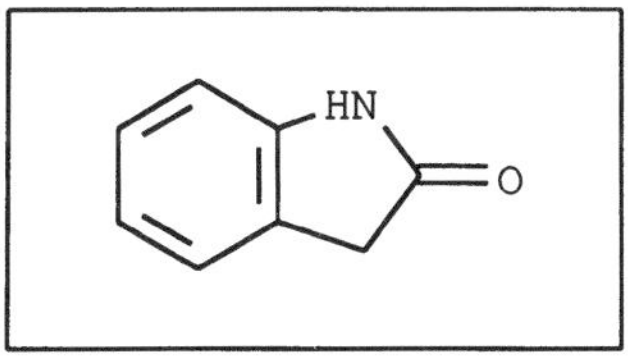

Autonom: 1,3-Dihydro-indol-2-one

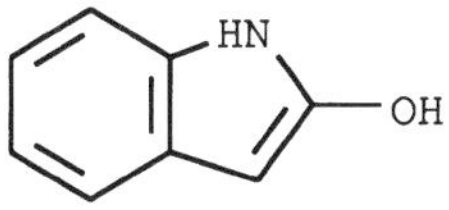

Autonom: 1H-Indol-2-ol

Figure 10.3. Handling of tautomers by AutoNom.

structures constitute the *complete and only* input information needed by the system's algorithm to generate the name.

It was decided that the system will generate names for explicitly entered bonds at the positions strictly determined during entering the structure (putting it succinctly: "what you draw is what you get"). Thus, both structures in Figure 10.3 should have the same name, whereas AutoNom generates two different names, ignoring the preferred tautomeric form (the one in the box).

The setup phase is formally complete after the atom connection matrix, atomic number vector, and charge vector are filled up with data. They are then submitted to the naming algorithm and processed during the naming cycle. The naming cycle is exhausted after a character string representing the chemical name is generated or a diagnostic message (informing the user that for some reason the program refrains from generating a name) is issued. It would be difficult to describe the exact flow of the processing for a computer program consisting of more than 51,000 lines of code divided into 460 various routines and functions (as profiled for version 2.0 in May 1996). However, one can select certain logical phases, understood as a sequence of conceptually related operations and tasks, that must be, one by one, graduated in the course of naming in order to deliver the chemical name textual string. The chronological sequence of the naming phases of the input structure is illustrated in Figure 10.4.

This division also reflects the general approach adopted for the design of the algorithm, and this is based on the concept of localizing (and possibly selecting and fully describing) the smallest structural entities present in the structures. These entities are nothing more than skeletal units, that is, catenated or cyclically connected atoms forming discrete nameable objects such as functional groups, chains, or ring systems. The

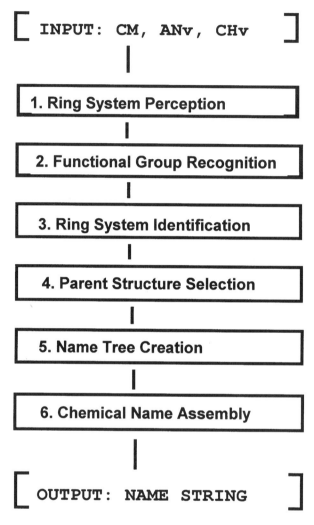

Figure 10.4. Phases of algorithmic naming.

selection and proper identification of the objects take place during the first three phases of the naming cycle. On completion of the third phase, all atoms of the input structure are uniquely distributed among the identified objects. So far, no hierarchy among the objects has been established (no seniority differences among objects are distinguished yet). The hierarchy is then determined using nomenclature rules during the parent structure selection and name tree creation. The descriptive term *"name tree"* used here is by no means abstract. The objects and their mutual relations are, on completion of the fifth phase, represented by a computer data structure having a form of a tree with the parent structure as the root and substituents as branches of the tree.

During the last phase of the cycle, hierarchically ordered objects (functional groups, chains, and ring systems) are related to the corresponding text fragments (determined by the syntax rules of IUPAC nomenclature), which are then compiled using appropriate IUPAC recommendations into the complete name of the input structure.

Ring System Perception

Prior to development of the AutoNom system, preliminary investigations of the possibility of algorithmic name generation had concluded that the most difficult problems would lie in the area of naming ring systems and assigning ring locants. It became obvious from the beginning that the reliability of ring identification is the crux of the naming algorithm.

The complete identification of any ring system present in the input structure is accomplished as a two-step process. First, the ring system must be perceived or detected, which means that all cyclic closures within the smallest sequence of atoms must be found—and related to appropriate single ring systems, if there are many. Second, the perceived ring system must be identified, that is, its name must be constructed or looked up in a dictionary and the right numbering must be resolved. The perception of ring closures present in the input structure and their classification in terms of nomenclature demands are crucial for successful identification and will be described in detail here.

Rings and ring systems are acknowledged to be the most important feature of the majority of organic structures, and their overall importance is not reflected any better than in conventional nomenclature.[48] Most ring systems are simply isolated rings or paired fusions; the task of finding these rings is trivially simple. What seems to be important is to be able to treat even the most complex cases in the same way. Complex cases will take considerably longer in machine-processing terms, but the results must be consistent with analyses of simpler ring systems.

Ring perception is a routine task, a utility to provide a description of cyclic parts of the structure in such a way as to complement other topological analyses, such as functional group recognition and chain fragmentation. The nature or perception and the set of rings to be found are application dependent. For IUPAC nomenclature purposes, the most important factors are topological interrelations between the smallest perceptible units in the ring system, reflecting the need to reconstruct its internal composition using the most discrete consistent cyclic entities and relate them with alphanumeric descriptors. In order to provide the obligatory unambiguous results, the set of such discrete cyclic entities should be as small as possible.

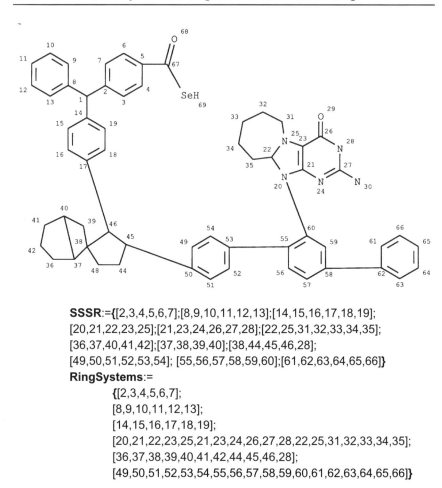

SSSR:={[2,3,4,5,6,7];[8,9,10,11,12,13];[14,15,16,17,18,19];
[20,21,22,23,25];[21,23,24,26,27,28];[22,25,31,32,33,34,35];
[36,37,40,41,42];[37,38,39,40];[38,44,45,46,28];
[49,50,51,52,53,54]; [55,56,57,58,59,60];[61,62,63,64,65,66]}
RingSystems:=
{[2,3,4,5,6,7];
[8,9,10,11,12,13];
[14,15,16,17,18,19];
[20,21,22,23,25,21,23,24,26,27,28,22,25,31,32,33,34,35];
[36,37,38,39,40,41,42,44,45,46,28];
[49,50,51,52,53,54,55,56,57,58,59,60,61,62,63,64,65,66]}

Figure 10.5. The concept of SSSR and the transition to nameable ring systems.

The model of perception that fulfills these needs can be described as the Smallest Set of Smallest Rings (SSSR).[49] It was never categorically stated or reported in the literature, but it was always intuitively clear that nomenclature guidelines for naming cyclic systems are based on the SSSR. The von Baeyer systems, spiro system nomenclature, etc., name and assign locants after first splitting a cyclic aggregate into the smallest set of the smallest rings in terms defined above. It was obvious that the AutoNom system would adopt the SSSR model for its ring perception routine. Figure 10.5 illustrates the transition from the perceived SSSR into ring systems for a given structure.

It can be seen from Figure 10.5 that the concept of "a ring system", in terms of nomenclature, describes the system as the unity of all cycles that share at least one atom with one or more other cycles in the unity. It has, as is usual in systematic nomenclature, an exception. Three phenyl rings connected to the common methyl group (Atom No. 1) constitute three separate, nameable ring systems, whereas the same three phenyl rings linearly connected to one another (Atom No. 49 through Atom No. 66) constitute one ring system, a so-called ring assembly, that must be named as a single cyclic unit.

There are many algorithms implementing the SSSR ring perception.[49] They are mostly based on graph theory and implement more or less the methods used in tree traversals and tree searching. For the purpose of AutoNom, the author has designed and implemented his own SSSR algorithm. It avoids the problems that result from the generation of "all" rings, where the large number of rings that is sometimes involved can lead to an exponential increase in the time required to generate them; for an interactive "draw and name" computer system, that increase could lead to unacceptable response time degradation.

Once the SSSR for an input structure has been generated, the transition from the detected SSSR to the nameable ring systems is accomplished by a set of dedicated, simple routines. Their main task is to localize the atoms shared by the cycles from the SSSR and compact the cycles sharing these atoms into one ring system.

Much more complex routines of the same type have been written to determine which ring systems constitute ring assemblies in nomenclature terms (linear composition of two or more identical ring systems joined by acyclic single or double bonds, not necessarily at equivalent positions). The decision to identify a cyclic system as an assembly is made in this early phase of the algorithm and influences further treatment of the assembly in the complete hierarchical analysis during the seniority comparisons.

Once all the ring systems, including assemblies, have been localized and their atoms marked, AutoNom generates, for each ring system separately, two additional pieces of information: (1) an array of pointers to subrings, and (2) hash code. These will later facilitate the task of final ring description during the recognition phase.

Many ring system classes, in order to be named, must be fully described by the specifications of all subrings constituting the system. This is the case, for example, for von Baeyer polycycles, spiro connected polycycles, and some classes of hybrid bridged and fused aromatic systems, to name only a few. This is one of the reasons why the information offered by the SSSR is valuable and as such must be stored for further use. For each ring system in the input structure, AutoNom produces the list of pointers to all cycles (represented as atom numbers involved in a ring closure, as illustrated in Figure 10.5) listed in the SSSR and constituting the particular ring system. This way, the internal composition of all subrings in a given ring system is always available to the algorithm. Together with the connectivity matrix (CM), it is more than enough for some ring systems (e.g., von Baeyer polycycles), to be fully identified and completely named and numbered.

In order to optimize and drastically shorten the time spent on searching ring systems whose names cannot be generated algorithmically and which must be looked up in a dedicated dictionary, the idea of similarity of chemical structure was recalled. The concept of similarity of ring systems may enable grouping of all ring systems bearing internal structural similarity into a single collection of rings always described with identical characteristics. Thus, any searching algorithm can always limit its searching to the same small collection of similar rings instead of the set of all rings. The best method of expressing this similarity in terms of a single descriptor, derived strictly from the CM and atomic number vector, AN_v, is the so-called hash coding. The hash code, which is nothing more than a short string of numbers or a single number, represents, in compact form, information on the atom characteristics and atom interconnections of a given ring system.

The simple, nonreversible hashing scheme algorithm has been developed by the author exclusively for use by the AutoNom system. It converts the CM and atomic number vector, AN_v, of atoms in a given ring system into an 18-byte-long hash string of positive integers. The resulting hash code is unique in that the same ring system, having the same number and art of hetero atoms and the same number of subrings located in a given relation to one another, will always be tagged with the same hash label. It does not mean, however, that each ring system has a unique code (such a noncollision hashing algorithm[50] does not exist so far). All rings selected to be included in the ring dictionary are prescreened with the hashing algorithm, and ring systems possessing the same hash code are grouped in the same pot. Thus, all ring systems are distributed into different pots, and each pot can be easily addressed. All pots are then written, using carefully selected, compact representation, into one ring dictionary file delivered as part of the AutoNom system.

Recognition of Functional Groups

Functional group identification is the second major step in the algorithm. The routines and functions within this step use a table-driven approach to first recognize and then identify, and rank in seniority order, functional groups. Technically, the table-driven approach is based on lookup processing with an efficient and rapid atom-by-atom connectivity search mechanism[51] as encountered by substructure search methods. Acyclic portions of the input structure are browsed in order to localize skeletal units bearing functional group characteristics such as hetero atom arrangements with unsaturated bonds. Each localized skeletal unit that is a candidate functional group is compared, in an atom-by-atom manner, with entries of a predefined, ordered dictionary of functional groups. The functional group dictionary contains carefully selected entries in the form of structural units, stored in a format optimized for both atom-by-atom search speed and minimal storage demands. The entries in the dictionary are ranked and ordered according to IUPAC recommendations for characteristic groups in substitutive nomenclature as stated in Rules C-10.1 through C-12.8 (reference 24, p 85 ff).

Each entry in the dictionary has (1) its unique precedence tag (for priority), (2) a skeletal substructure unit associated with it (atoms and atom interconnections in the substructure are compared in the atom-by-atom search against the acyclic units selected from the input structure), (3) a detailed specification of the generic nature of its further attachments (if any), and (4) a textual descriptor of its suffix form (if the functional group can possibly be cited as a principal group) and obligatory prefix form. The skeletal substructure of any functional group selected as a dictionary entry is stored internally as a connection table in a compact format designed for the atom-by-atom procedure encoded into the algorithm.

This redundant representation, besides the normal CM, AN_v, and CH_v, also contains the backtracking list of all atoms up to four neighbor atoms forward and four neighbor atoms backward, and a specification of the preferred attachments (single atom, cyclic group, acyclic group, etc.). Although redundant, this information was found to be important to design a search algorithm that enables a drastically increased speed of atom-by-atom dictionary matching. It is eminently important because the functional group identification is a step of the naming algorithm that is conducted for any structure containing hetero atom characteristics, and these are in the absolute majority.

On completion of the functional group dictionary lookup, the units that were identical to entries in the dictionary are sorted in priority order (guided by nomenclature considerations mentioned before), and a ranked list of all groups present in the input structure is compiled. The highest-ranking groups are then used in the parent-structure selection phase (as discussed in detail later) to generate a set of candidate parent fragments. The term *highest ranking* is derived from the IUPAC recommendations[24] and is supposed to set clear conditions on the selection of the so-called *principal functional group*, which later (in the complete name), is cited as the suffix or is a functional parent (e.g., carbamothioic acid) rather than an ordinary group.

The task of functional group recognition, although here described as the sovereign phase of the naming cycle, is not usually accomplished in a single run. A list of functional groups ranked according to nomenclature guidelines is delivered for further processing. The entries on the list as well as data describing the entries normally do not stay in the initial form throughout the whole naming cycle. The initial form is, however, extremely important for selecting the principal group(s), which must be fully identified before the algorithm enters the parent selection phase.

Later, after the parent fragment of the input structure has been selected, the list is usually modified by processing it, at least once, with functional group updating routines.

On completion of functional group recognition, the first class of objects, that is, acyclic chains bearing hetero atoms and linked with chemical functionality, are fully selected from the input structure, hierarchically ordered or ranked, and preliminarily identified to the extent enabling their naming. Further phases of the naming cycle must identify the other two classes of objects, that is, ring systems and chains.

Ring System Recognition

The preliminary ring perception phase, as discussed earlier, delivers a complete list of all localized ring systems, including ring assemblies. Each ring system in the list is described by the set of atoms constituting the system (Figure 10.5), the array of pointers pointing to positions of its subrings in the previously generated SSSR, and by the 18-byte hash code. All the rings are supposed to be recognized with only this information, that is, their names must be generated and the correct numbering of their atoms must be set.

As far as ring system recognition is concerned, there is a dual approach implemented in AutoNom: (1) dictionary lookup with an atom-by-atom search mechanism, and (2) strictly algorithmic naming and numbering exclusively on the basis of the internal composition of the processed ring system. The policy adopted by the author during work on the AutoNom project was to cover the greatest possible number of ring system classes with algorithmic identification and refer to dictionary access only if the latter proved to be distinctly more efficient, or only possible from a system performance point of view. The classes of ring systems that are recognized purely by an algorithmic analysis—each class has a complex dedicated collection of routines and functions performing the task—include:

- monocyclic alkanes (e.g., cyclohepta-1,4-diene)
- bicyclic (hetero)alkanes (e.g., bicyclo[3.3.2]deca-1(8),2,5,9-tetraene)
- monospirocyclic (hetero)alkanes (e.g., spiro[4.5]deca-1,3,6,8-tetraene)
- dispirocyclic (hetero)alkanes (e.g., dispiro[4.2.6]hexadeca-1,3,6,9,12, 15-hexaene)
- heteromonocyclic named by the Hantzsch–Widman method[48] (e.g., 2,5-dihydro-2H-[1,3,7,5]oxabismaplumbaine).

The complete systematic name of any ring belonging to the above classes is constructed purely on the basis of information derived from the hash code, pointers to SSSR cycles, the CM, and AN_v.

A name of any ring system not belonging to one of the above-specified classes is generated after lookup dictionary access to trivial name ring systems. An atom-by-atom matching is conducted in an earlier prepared dictionary containing ring connection tables together with prescribed numbering and name strings. Much attention has been paid to optimizing the organization and constructing the dictionary in terms of efficient processing demands. It is obvious that the size of the dictionary will grow with each new update version of AutoNom. The initial rationale for choosing rings, mostly annelated aromatic fused systems, for the dictionary was based strictly on statistics of occurrence of these systems in organic compounds. Any ring system whose frequency in the Beilstein database was measured as exceeding 5% was selected for inclusion in the dictionary. In addition, for the updates of the program from version 1.0 to version 1.1 and later to version 2.0, all documented suggestions of important ring systems coming from users of AutoNom were fully acknowledged, and the dictionary was richly augmented. This practice will also be followed in all upcoming versions.

An important factor for a constantly expanding dictionary was minimization of storage space of the dictionary file at optimized access time to its entries. This was achieved by using the author's own customized data compression and decompression techniques[52] for storage and retrieval of entries from the dictionary file.

The search in the dictionary of rings differs only slightly from well-known substructure searching methods.[53] Instead of looking for a single substructure in the set of prescreened structures, a set of perceived cyclic substructures is sought in a single (input) structure. The whole process might, to coin a phrase, be described as an "infrastructure search". The query patterns are compared with all members from a dictionary pot (the beginning of the pot and its size are determined using the hash code) until an exact match is obtained. To speed up searching, only interring atom connections—with no information on bonding art—are stored in the CM of any dictionary entry. It eliminates, in this phase of the naming cycle, a bond restoration and saves time normally spent on bond denormalization and unscrambling during search.[51]

Having matched the ring system with a dictionary entry, the algorithm continues with the complex task of ring atom numbering. For nonsymmetrical systems, the situation is relatively simple. Each atom of a ring requires only one fixed locant, which means that for each dictionary entry only one collection of fixed locants needs to be stored with the entry. For symmetrical systems this is unfortunately not the case. Depending on the number of symmetry axes, any single atom in a ring can be numbered with many equivalent locants. It would mean, in practice, that not one but all possible combinations of prescribed fixed locants, for all allowed numbering schemes, should be stored with every symmetrical ring entry.

"1"   "1"

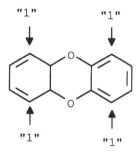

"1"   "1"

Figure 10.6. Alternatives by numbering of symmetrical ring systems.

In the above example of dibenzo[1,4]dioxine (Figure 10.6), each of the four marked atoms can, in principle, be given the number "1" locant.

Which of them is finally assigned "1" is decided in accordance with IUPAC recommendations C-15.1 and C-15.2 (reference 24, pp 105–107). Normally only this atom is enumerated as "1", which guarantees that the rest of the atoms as a complete set will get the lowest possible locants. While deciding on enumeration for such systems, the following criteria are applied until a decision is reached:

(a) lowest locants for principal groups(s) cited as a suffix,
(b) lowest locants for indicated hydrogen,
(c) lowest locants for multiple bonds in rings whose names indicate partial hydrogenation (cycloalkanes/enes/ynes, pyrazolines, and the like),
(d) lowest locants for substituents named as prefixes, and
(e) lowest locants for substituents named as prefixes in alphabetical order of citation.

However, in order to apply these criteria, the applicable combinations of fixed locants (the complete sequence of locants for all atoms in a ring) must be known beforehand. For the dibenzo[1,4]dioxine in Figure 10.6, there are only four such combinations, but for the extreme case of the highly symmetrical cubane, the number rises to 24. The idea of storing all these combinations as arrays of locants in the dictionary, even in the most compressed form, was rejected as absolutely infeasible from the point of view of both storage requirements and the estimated dramatic degradation in system performance and response time. Instead, a relatively simple routine based on look-ahead techniques has been designed by the author and included in the atom-by-atom search procedure. The routine is capable of generating all possible permutations of locant compositions strictly on the basis of one composition stored in the dictionary entry and the two-dimensional CM of the ring system that was matched with the hit entry. The small redundancy in search time (approximately

11.8% for an average system with eight possible combinations of its numbering) can be tolerated taking into account the savings in global storage of the dictionary and prolonged access to entries due to the increased size of a single entry. (Entries in a given ring pot are accessed sequentially until a total match is achieved or, in the extreme case, the whole pot has to be processed in case of no hit.)

There is also a relatively large group of ring systems for which a combined dictionary and algorithmic scheme of identification is desirable. This combined approach is applied for all fused polycyclic hydrocarbons and heterocarbons with monospiro or dispiro connections to a ring system classified as a dictionary entry.

Fine recognition of ring assemblies is the final step in the ring recognition phase. During this phase, it consists of identifying all isolated ring systems involved in an assembly union and treating them accordingly, that is, as being a part of a bigger cyclic unit. Usually it means additional complexity, as, for example, including specific nomenclature criteria such as a special status for the point-of-attachment atoms in the process of numbering or assigning nonprimed, primed, or multiprimed locants to atoms constituting particular ring systems of the assembly.

The recognition of all ring systems for most of the structures usually involves the application of both methods: algorithmic recognition and lookup atom-by-atom search identification. In extreme cases, even for a single structure a great variety of different algorithms must be used (and the job of the system is to decide which rings are identified by which method) to name and assign locants to atoms in all the rings in the input structure. How complex the whole procedure can be is well illustrated for a rather more hypothetical than realistic structure in Figure 10.7. It contains seven ring systems representing seven different ring classes. For each ring class, AutoNom has a set of special routines and functions that must be invoked in order to generate the ring's name and set up the numbering of its atoms. Except for ring system R1, all ring systems are identified strictly algorithmically, without consulting the ring dictionary. The name "adamantane" for the tricyclic von Baeyer system R5 is derived from its systematic equivalent tricyclo[3.3.1.1^{3,7}]decane in a simple string replacement operation. It is also valid for all Hantzsch–Widman five-membered thioles in R6, later changed into, retained, and fully sanctioned by IUPAC as "thiophenes". The three thiophene rings are identified separately at first and then determined to constitute an assembly, and they are renamed the assembly name accordingly.

Only dictionary ring system R1, (fused aromatic) benzo[c]chrysene, is searched in its corresponding pot, whose address, relative to the beginning of the dictionary (and size too), is derived directly from its 18-byte hash code.

After identifying all the rings, the most senior one, in this case the parent unit, must be selected. Selection is accomplished by a set (39

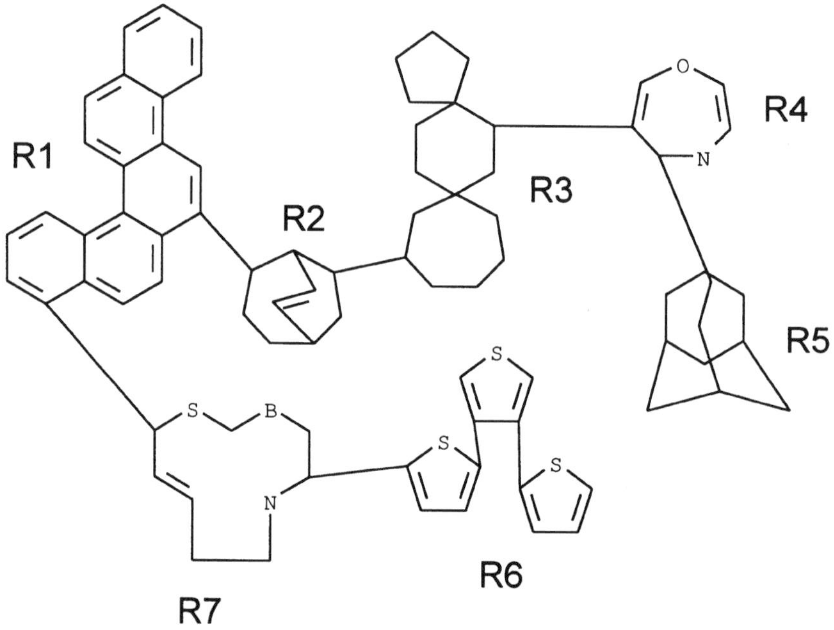

AutoNom: 11-(6-{8-[6-(5-Adamantan-1-yl-4,5-dihydro-[1,4]oxazepin-6-yl)-dispiro [4.2. 6.2]hexadec-10-yl]-bicyclo[3.2.2]non-6-en-2-yl}-benzo[c]chrysen-9-yl)-5-[2,3(,4(,2((]terthiophen-5-yl-1-thia-6-aza-3-bora-cycloundecane

Figure 10.7. Variety of methods based on identification of different ring system classes.

various routines and functions) of dedicated procedures during the parent structure selection phase. Seniority here is uniquely decided by the greatest number (and variety) of hetero atoms occurring in a ring system (in full accordance with rules A-14.11 and B-3.1 in the *Blue Book*). Thus, the "a" replacement mono heterocyclic R7 is chosen as the most senior ring and, simultaneously, as the parent structure of the compound in Figure 10.7.*

Selection of the Parent Structure

The generation of a set of candidate parent structures (for the stem structural unit of the input structure) is the next important task performed

---

* The complete name for this rather exotic constellation of rings in a single structure is generated by AutoNom within approximately 25 s (for AutoNom 1.1 and a Compaq 80486/66-Mhz personal computer). This relatively long time (for a highly skilled nomenclature specialist from Beilstein it took ca. 10 min to name this compound, that is, approximately 22 times as long!) is due to the dictionary search of the benzo[c]chrysene (its corresponding pot contains a great many ring systems) and the ring assembly (terthiophene) processing.

during the naming cycle. Once the functional groups and ring systems have been recognized, the algorithm proceeds with the identification of chains. They are, besides functional groups and ring systems, the third and simultaneously the last category of objects that must be identified in the input structure in order to divide it into parts hierarchically related to each other.

The selected potential candidate structural fragments (ring(s), chain(s), or, in extreme cases, functional groups, but only if the name of the structure is built around a so-called functional parent) are then ranked according to the relevant nomenclature principles. If more than one candidate structural unit of the same rank competes for selection as the parent, the following global sequence of principles, in the order given, is applied until a decision is reached:

I.   The greatest number of principal functional groups cited as suffix. The rationale adopted here is identical to that of IUPAC as stated in C-12.1 through C-12.6 (reference 24, pp 92–96). In case of more than two objects carrying the same number of principal functions, IUPAC is not very specific. Should these two objects be chains, then the normal criteria for seniority of chains are applied. The same is valid for ring systems, except that the seniority criteria for rings must be used.

II. Preferred hetero content of the molecular skeleton. Rings are preferred to chains, and within rings the hetero content is ranked according to the global variety of hetero atoms as well as the predefined priority order of the hetero atoms.

III. Seniority of rings (if only ring candidates are left).

IV. Seniority of chains (if only chain candidates are left).

V.   The greatest number of multiple bonds.

VI. The lowest locants for successive hetero atoms, principal groups (suffixes), all multiple bonds, and all double bonds.

VII. The greatest number of substituents.

VIII. The lowest locants for the substituents cited first as prefixes in alphabetical order.

The structural fragments that cannot be assigned the status of parent automatically become substituents on the selected parent structural unit.

It is worth noting here that upon leaving the parent structure selection phase, except for atoms belonging to substituent chains (and only those chains not being considered as potential parent candidates), all atoms of the input structure are hierarchically distributed among nameable structural units.

The above eight general principles of parent structure selection are the most important criteria in the determination of the seniority relations in the input structure by AutoNom. They obviously do not represent all the rules implemented in the system, but they are the most relevant.

A particular group of selection criteria, which frequently alone lead to the unique selection of the parent structure, are the rules determining seniority of chains and seniority of rings. The seniority of ring systems is decided by applying 39 criteria as used by the Beilstein Nomenclature Department. They are, in general, in agreement with the corresponding IUPAC recommendation, C-14.1 (reference 24, pp 101–105). The principal chain, that is, the chain upon which the nomenclature and numbering of the whole structure will be based, is chosen by successively applying 10 criteria formulated by IUPAC recommendation C-13.1 (reference 24, pp 97–100).

Creation of the Name Tree

Preliminary investigations prior to the development of AutoNom had concluded that the hierarchic principle underlying the approach to chemical name construction (parent, substituent, substituent-on-substituent, etc.) should be followed as faithfully as possible during design of the appropriate data format for name generation analysis. It was decided to implement a format based on an ordered binary tree concept[54] as fulfilling the majority of both nomenclature and system-performance requirements. The data structure maintained in the memory of the computer during nomenclature-guided analysis of the input compound will be hereafter referred to as the *name tree*.

Each node of the name tree corresponds to one nameable unit of the input structure and contains a *record of data* characterizing the unit. The parent structure fragment, selected in the preceding phase of the algorithm, is established as the root of the name tree, whereas the other nameable units are so-called substituent nodes of the tree. The name tree data structure and the corresponding compound are illustrated in Figure 10.8.

The base node (the root of the tree) describes the parent molecular skeleton—ring system, chain, or functional parent—whereas the other nodes represent the other nameable units localized and identified in the input structure. Starting at the root and traversing in an upward direction, mutual relationships among the nodes (e.g., type of bonding, locants of attachments, indicated hydrogen locants for ring systems, locants of multiple bonds in chains, etc.) are sought, and the complete data on this relationship is added to the existing node records or a new record is initiated, with these data, if none yet exists for this node. Concurrently, if necessary, existing nodes are removed from the tree or brand new nodes are formed and added to the tree.

The mutual relations among nodes and the hierarchical sequence of subnodes for a given node are maintained by a system of pointers and so-called node tags represented as variable-length strings of characters

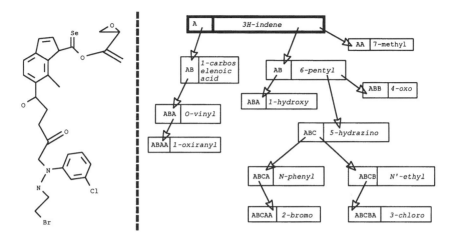

STRUCTURE   **NAME TREE**
AutoNom: 6-{5-[*N*(-(2-Bromo-ethyl)-*N*-(3-chloro-phenyl)-hydrazino]-1-hydroxy-4-oxo-pentyl}-7-methyl-3*H*-indene-1-carboselenoic acid *O*-(1-oxiranyl-vinyl) ester

Figure 10.8. Construction of the name tree derived for a given input structure.

(e.g., ABAA in Figure 10.8). Knowing the tag of a given node, the whole route from that node to the root of the name tree can be easily reproduced at will.

The design of the algorithm assures that the information stored in data records describing the particular nodes of the name tree is the "best" possible at the moment, but only at *the* moment, and it does not mean, however, that it is good enough to construct the chemical name. For certain ring classes, their name descriptors are not yet fully complete. For example, "cyclopropa[3,4]pentaleno[1,2-*d*]dioxole" is stored in the record of the appropriate node instead of the complete form, "2a,2b,2c,5a,5b,5c-hexahydro-cyclopropa[3,4]pentaleno[1,2-*d*]dioxole". The missing hydrogenation* prefix will be algorithmically generated in the next phase of the algorithm. The correct generation of the "2a,2b,2c,5a,5b,5c-hexa-hydro" can only be successful after the complete set of all subnodes of the cyclopropa[3,4]pentaleno[1,2-*d*]dioxole node is identified and described. Functional groups are still described with both their suffix and prefix forms (e.g., "carboselenoic acid/selenocarboxy"). In the next phase of the naming cycle (name assembly), first the algorithm will decide, by examining the principal or nonprincipal status flag of the functional group, whether

---

* Nota bene: The hydrogenation routine was among the most complex tasks of the whole AutoNom project in terms of design and programming, and it is also the most praised, by the users of AutoNom, for utility of ring analysis. For many fused polycycles, finding the best locants for indicated hydrogens and positions of hydrogenation is much too complex for a practicing chemist.

it should be "selenocarboxy" as a prefix or "selenoic acid" (for an aliphatic parent) or "carboselenoic acid" (for a cyclic parent) as a suffix.

Name Assembly

The construction of the entire name tree of the input structure is now the starting point for final launching of the name assembly phase of the algorithm. The assembly routines build a name for the name tree given, determining the appropriate name fragment for each node in the tree and combining these fragments with proper ordering, locants, and punctuation. Beginning at the highest so-called terminal node (a node which is topologically most distant from the root, that is, parent skeleton, of the tree) and traversing the name tree downward, the data records of visited nodes are processed and the resulting textual name fragments (e.g., 1-carboselenoic acid, 7-methyl, 5-hydrazino, etc., for the tree in Figure 10.8) are stored in a so-called *name fragment table*. In order to keep track of the path and the sequential order of the nodes visited while traveling from a given node to the root of the tree, the tags of the nodes are immediately stored on a dedicated reversed stack. Thus, the first pass of the name assembly has been completed.

During the second pass, starting at the top element of the tag stack, the corresponding name fragment entries, pointed at by the tags, are successively extracted from the name fragment table and combined—by applying the proper chemical nomenclature semantics and syntax—into longer fragments. The whole process repeats itself until the tag stack is emptied (the root node was reached) and the first version of a complete name is obtained. The job of combining many shorter fragments into one longer name fragment, and finally, in conclusion, into the final name, is performed by several elaborate routines that handle such sophisticated operations as alphabetization, multiplication, punctuation, vowel deletion, suppression of unnecessary locants, superscript and italic string placement, etc.

These routines are by no means trivial. Even such obvious operations (for humans) as the suppression of unnecessary locants is a surprisingly complex task. Because of the global logic of IUPAC nomenclature, AutoNom always generates locants even for such obvious and trivial cases as 3-(1-chloromethyl)-indene, and then first suppresses them, but only if necessary.* Whether the condition "if necessary" applies is the clue to the whole procedure. It must be explicitly stated in the algorithm when locants are to be cited and where allowing locants to be omitted would not cause ambiguity in the resulting name. This task is handled by a dedicated set of routines, and these are complex because they must usually

---

* The locant suppression has further consequences here: the name fragment *(1-chloro-methyl)* is replaced by *-(chloromethyl)* and as such does not need parentheses. AutoNom has to be able to "notice" that and remove them; the final name is then *3-chloromethyl-indene*.

determine whether the substructure under consideration is symmetric in some respect. The presence of symmetry is a general indication that a locant may not be needed. Concluding that a structure skeleton is symmetric always means a complete analysis of the whole structure, not only that of the skeleton considered. Sometimes even symmetry is not the real reason why chemists (IUPAC, unfortunately, sanctions these practices) prefer names with no locants. They find, in opposition to many nomenclature chemists and practically all nomenclature software developers, that citing them in a name is overredundant and that their lack is "obvious" and "normal" and makes the name look more "intelligent". In practice, it means, for example, that for the compound

AutoNom: 2-Chloro-*N*-propyl-acetic acid amide

citation of the locant for the "chloro" substituent is the normal practice (because of the substituent on the nitrogen of the amide), whereas for the esterified acetic acid in a very similar compound,

AutoNom: Chloro-acetic acid propyl ester

citing the locant is absolutely redundant. AutoNom has to be "intelligent" enough to differentiate between such cases and react appropriately.

The name fragment terms used by the name assembly procedure are derived from the analysis data records corresponding to the particular nodes (the records contain information necessary to build the textual descriptors, but not descriptors themselves). The information so far completed in the records is processed and augmented in order to construct the full name fragment of a given object represented by the node:

1. *Ring names.* Descriptors of algorithmic ring classes are derived from the records describing internal composition, hetero atoms occurring in the ring, and unsaturation characteristics. Indicated hydrogen or hydrogenation locants are set on the basis of all possible permutations of allowed numbering compositions.
2. *Chain names.* Chain characteristics such as length, position of multiple bonds, and position of substitutions are extracted from the corresponding records and transformed into chain descriptors. Selected routines and functions handle special cases of trivial or semitrivial names retained and sanctioned by IUPAC, such as allyl or vinyl.
3. *Functional group names.* Those which can be split no further are directly derived from their suffix + prefix descriptors stored in records (and coming from functional group tables). It must only be decided which form, suffix or prefix, to extract. For groups that must be split first, the splitting operation takes place, the name tree is updated, and the appropriate descriptors (always the prefix form) are simply overtaken.

The generation of name fragments from data records of currently visited nodes of the name tree, and the combining of these fragments into bigger name units, is, from the beginning, monitored without interruption by intelligent (so-called "triviality") controller routines. The controller is responsible for tracking—and immediately replacing—any "too systematic" and thus not intelligible nomenclature with well-established, IUPAC-sanctioned, nonsystematic, traditional nomenclature. The question of whether something should ever be replaced and what is "too systematic" and should be replaced is a subject of heated and long debates in the chemical literature[55] and by IUPAC itself within the Commission on the Nomenclature of Organic Chemistry[56]. To illustrate the issue, the following simple structure can be considered:

AutoNom: Benzoic acid

As for "benzoic acid", this name is obviously not formed according to either of the main principles[24] for forming systematic names for carboxylic acids, namely, (1) adding "-oic acid" to the name of the hydrocarbon having the same skeleton (in this case, it would be toluene, also a trivial name, as a matter of fact), or (2) adding "-carboxylic acid" to the name of the parent hydride in which H is replaced by –C(=O)OH (i. e., benzene). Naming acids systematically was already a vexing problem for the participants in the conference in Geneva in 1892, and it continued to

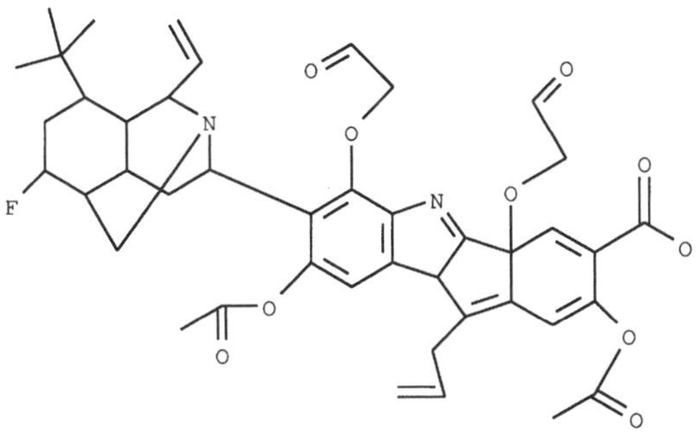

AutoNom: 2,8-**Diacetoxy**-10-**allyl**-7-(6-*tert*-**butyl**-8-fluoro-9-**vinyl** -octahydro-

2,5- ethano-isoquinolin-3-yl)-4a,6-bis-(2-oxo-ethoxy)-4a,10a-dihydro-indeno[1,2-

b]indole-3-carboxylic acid

Figure 10.9. Trivial (IUPAC sanctioned) nomenclature in names generated by AutoNom.

be difficult long afterwards. The best that can be done in the framework of current IUPAC rules is "benzene-carboxylic acid" (or perhaps "phenyl-formic acid"), but neither of these is actually recommended by IUPAC. One can certainly deduce the structure unambiguously from these names (if one accepts their trivial components "benzene" or "phenyl"), but they are considerably longer than "benzoic", and for such a commonly mentioned compound, that is a serious disadvantage.

This discussion on the subject of where "systematization" starts can go further. A few militant prolix users of nomenclature would prefer that the structure have the bizarre name "cyclohexa-1,3,5-triene-carboxylic acid", which is fully, and perversely, systematic, and they have many good arguments that their approach is the only correct one because it is the only one that is systematic in every detail. No matter how tempting and elegant the idea is from a computer's point of view, AutoNom cannot go so far with systematization, for if it did, no copy of this program would have a chance of being sold and used.

Incorporation of trivial names into a consistently systematic computer algorithm is by no means simple and straightforward, as shown by the analysis for the structure in Figure 10.9. The transient prefix name fragment *(1-oxo-ethoxy)* has been replaced with *acetoxy*, which is preferred by Beilstein.

On the other hand, the prefix *(2-oxo-ethoxy)*—and the triviality controller should "know" this—does not need any further changes. The simple string-for-string replacement in the first case has important consequences as far as the syntax of the global name is concerned. The name fragment *(1-oxo-ethoxy)* would be multiplied with the multiplication affixes bis-, tris-, tetrakis-, etc., as approved for condensed substituents. Deciding on this multiplication scheme means that during alphabetical ordering of prefixes, *diacetoxy,* according to C-16.31 (reference 24, p 109) as starting with an *a,* is a so-called simple prefix (to make it even more complicated, the multiplying affix *di-* does not count according to the same rule), and it is cited as the first in the row of all prefixes, before *allyl* as having an *l* as the second letter in comparison to *c* in the acetoxy. On the other hand, the *(2-oxo-ethoxy)* prefix, which differs from *(1-oxo-ethoxy)* → *acetoxy* only by the locant of the oxo group, should be a condensed prefix multiplied with the *bis-* affix. In addition, it stays enclosed in parentheses and starts with *o* (in *oxo*), and this is why, in the complete name, it is cited as the last prefix (the nondetachable "4a,10a-dihydro-" is always cited in front of the ring system and does not influence the alphabetic ordering of detachable prefixes). The other trivial parts that are encountered in the name and dealt with accordingly are "allyl", "*tert*-butyl", and "vinyl". The additional complexity of *tert*-butyl consists in special treatment of the *tert-* subprefix. It must be implemented in the algorithm that any italic subprefixes (*tert-, sec-, o-,* etc.) not be treated as part of the name fragment and that they be fully ignored by alphabetization. As in almost everything in systematic nomenclature, this rule also has exceptions: in the case of "*tert*-butyl" and "*sec*-butyl" competing for citation, "*sec*-butyl" is chosen as being alphabetically senior (even though, as a rule, the italic subprefixes should be ignored) because *sec-* is alphabetically senior to *tert-*.

Finally, there are a few operations that are not necessary for name generation but that are useful and sometimes appreciated by the users. These include (1) capitalizing the first relevant letter of the name, and (2) changing parentheses, according to a detailed and approved schedule, into sequences of (, [, {, },],) closed pairs. Even these operations are not that simple; AutoNom must "know" that in the name 1-([3-{2-*tert*-Butyl ...), the first relevant character is B (in Butyl) and that the preceding italic *tert-* must simply be ignored.

The triviality controller text string replacement described here should not be confused with another string exchange that can (if required) be implemented as a much simpler single-run process in a user-defined shell. This may be done at the very last stage of the name assembly, where, for example, natural language and usage considerations should be taken into account (e.g., 2-Amino-1,9-dihydro-purin-6-one → *guanine;* or hexanetrioic acid → *kwas heksanotrikarboksylowy* (Polish), etc.).

Nomenclature Implemented in the AutoNom System

According to IUPAC (reference 24, p 112), it is a fundamental rule that substitutive nomenclature is, in general, to be preferred. This is why the AutoNom naming system is designed and built around the principles of substitutive nomenclature, with a limited application of replacement nomenclature (only for heterocycles and not for chains), additive nomenclature (hydrogenation prefixes and cation and anion suffixes), and subtractive nomenclature (unsaturation in aliphatic compounds and in cycloalkanes, bicycloalkanes, etc.). The use of radicofunctional and conjunctive nomenclature is not supported, and, at present, this also applies to sometimes very useful multiplicative nomenclature.

The nomenclature generated by the AutoNom program follows "pure IUPAC" very closely indeed, although nomenclature experts will immediately recognize in the following the elements of the dialect of IUPAC used in Beilstein. On the other hand, the systematic approach had to be kept in frames set by the chemical community, particularly in the area of trivial or semitrivial names used in organic nomenclature.

There is a body of trivial names that are part of the general language and are far from being exclusive to chemists. "Sulfuric acid" is most unlikely to be supplanted by a coordination name for general use. "Iron" is certainly not going to be given up in favor of "duohexenium" or even "ferrum". "Sulfite", "bicarbonate", "acetone", "formaldehyde", "toluene", "styrene", and even "naphthalene", among a host of others, are firmly entrenched in the general language. An attempt to do away with such names only because their triviality is in collision with the hard logic of a computer algorithm would, at this time, result only in alienating the already suspicious public. This public would see the introduction of systematic substitutions as unnecessary and arcane, a means of excluding the layperson. The current IUPAC rules also recognize this by including statements such as "the name acetic acid is retained". The intent of this statement is clear: "acetic acid" is the recommended name, rather than the systematic "ethanoic acid" (or perhaps "dicarbanoic acid"). If AutoNom is supposed to retain an anchor to the real usage of organic nomenclature, it has to be able to generate names such as "acetic acid", "propionic acid" instead of "propanoic acid", and "butyric acid" instead of "butanoic acid". This had to be acknowledged—with disappointment, because it always means drifting away from the algorithmic logic path—at the beginning of the AutoNom project.

Performance of the Algorithm

The program's performance has been evaluated internally in Beilstein on multiple occasions. These evaluations produced important data on how Beilstein should be corrected and updated and in which direction further development should go. Besides this, an independent evaluation[15] (for Version 1.0) was conducted in Wyeth Ayerst Research (Princeton, NJ), where AutoNom was implemented (and since then regularly used) just after its release in 1991.

The version tested was the newest release of the program (Version 2.0). A sample selected for evaluation was a collection of 63,040 compounds randomly chosen from the Beilstein data bank. No special classes, types, or groups of organic compounds were preferred; each hundredth structure out of 6,349,225 active structures in the Beilstein database was simply transferred to the sample file, and AutoNom was run over the file. The program generated 54,088 names, which gives the general success rate (measured by the production of a name on a nonsteric basis) of 85.8%. In the remaining 14.2% of cases, the program refused to generate a name and delivered a diagnostic message explaining the reason why the naming could not be completed successfully.

The degree of acceptability of the names produced was measured by determining the so-called expert status of the names, where expert status was defined as output identical to that of the Nomenclature Department of the Beilstein Institute. In order to do this, a random sample of 2704 names (every twentieth name from the output file) was selected and checked for its correctness by the nomenclature specialists at Beilstein. The number of names not conforming to the expert status of the Beilstein Institute was 11, which gives a level of total unacceptability of less than 0.4%.

The 11 rejected names were fully correct as far as IUPAC rules of systematic nomenclature are concerned; they simply did not conform to the nomenclature dialect used by Beilstein. Using them as input to the VICA program, which does just the opposite of AutoNom (that is, it translates systematic nomenclature into structural diagrams), correct structures were received as output. Redirecting AutoNom's output (names) as input to the VICA program is the fastest (and also the least expensive) method of monitoring the quality of the names generated by the AutoNom program. The output of the VICA program (structures) is then compared, by a dedicated computer program, with the input of AutoNom. Both sets of structures should be identical.*

The evaluation test was run on a PC with a 66-MHz processor, and the average naming rate (for the whole sample of 63,040 structures) was measured at 35.6 structures per minute. It is obvious that the rate depends heavily on the complexity of the structures processed, but taking into account that the sample prepared for the testing was random, it gives a good estimate of the expected naming rate of the program.

In the case of the 8,952 rejected structures from the analyzed sample (14.2%), AutoNom did not generate any name producing diagnostic messages. Particular messages (known as Error Messages in the program) displayed, instead of the chemical name, a detailed reason for abandoning the naming. The messages usually refer to structural properties of compounds

---

* For many users of AutoNom, the only objective test of AutoNom's naming abilities is to take a structure, process it with AutoNom, redirect the output name as input to the VICA program, and obtain the identical resulting structure as VICA output.

that, for the current version of the AutoNom algorithm, are too complex. These properties cannot yet be correctly analyzed and processed by the algorithm because the corresponding routines and functions are not yet implemented These properties can be grouped into classes, and for the sample tested (8,952 diagnostic messages), the following statistics were obtained:

1. *Unidentified functional group [37.3%].* A functional group is an arrangement of hetero atoms with saturated and/or unsaturated bonds, and these are recognized by the algorithm using an internal dictionary of several hundred such atom compositions. Because the number of such characteristic groups is practically unlimited, AutoNom supports only the chemically and statistically most important ones. It can happen that certain groups are not represented in the dictionary, and then the above message is issued by the program. The dictionary of the functional groups can be augmented at will. When necessary (and reasonable), the next versions of AutoNom can be configured with a bigger dictionary of characteristic groups.

2. *Unrecognized ring system(s) [28.1%].* The problem of ring perception and ring recognition has been discussed in detail. AutoNom works with two classes of ring systems: rings whose names (and numbering) are generated strictly on the basis of their internal composition, and rings whose characteristics are looked up in the dedicated dictionary. In the case of the former class, it can happen that a ring system does not fall into the categories covered by the algorithmic methods. This is, for example, the case for tetracyclic (and bigger) von Baeyer ring systems. They are not handled by AutoNom; the frequency of their appearance in the Beilstein database lies below a fraction of 0.1 percent, and it makes no sense to implement complex and bulky routines in the program. For the latter, it can happen that a ring system is not (yet) included in the dictionary. Each upcoming upgrade of AutoNom will be released with a bigger dictionary, so the number of rejections due to "unrecognized ring system(s)" should drop gradually.

3. *Ring seniority problem [15.2%].* A structure for which this diagnostic message is issued contains at least two ring systems competing for seniority. The program, while choosing the senior ring, has reached the end of its present selection criteria and cannot decide which system should be selected. There are cases, usually for highly symmetrical structures, where applying all these criteria does not lead to a decision. This can happen even for simple compounds, such as tetraphenylethene. The choice of ethene as the stem of the name is strictly a common sense decision and not a clear application of the 39 rules, and this is obviously beyond any computer program capabilities.

4. *Charges too complex [3.1%].* The program can handle the following classes of organic cations and organic anions:

a. A positive charge on a single hetero atom in a ring system, as in 3-(2-anthracene-9-yl-vinyl)-5-phenyl-[1,2]dithiolylium, as well as a single positive charge on hetero acyclic units such as ammonium, sulfonium, diazonium, etc.

b. A negative charge on a single chalcogen atom in an acyclic unit, such as "acylates" and "alcoholates".

c. Positive or negative charge on one component (valid for two-component structures only) must be balanced by negative and positive charge on the other. (Only presence counts; the number of charges is not important.) Any constellation of charges that does not conform with this rule causes AutoNom to issue this diagnostic message.

5. *Atom(s) out of range [3.0%].* The set of heavy atoms for an organic component (hydrogen is implicit) is confined to those in Table I, and these may be used in any part of the structure. However, when used in chains, they must be part of recognizable characteristic groups. There are no limitations on atoms as far as inorganic components are concerned (provided each atom is a valid periodic table entry). Any use of invalid atoms in the input structure causes AutoNom to refrain from naming and to issue this diagnostic message.

Table I. Atoms Allowed in the Organic Component of the Input Structure

	B	C	N	O	F	
			Si	P	S	Cl
			Ge	As	Se	Br
			Sn	Sb	Te	I
Hg			Pb	Bi		

6. *Other diagnostic messages.* There are altogether 27 other diagnostic messages that can be issued by the AutoNom program, and they refer to various structural characteristics or computer environment limitations that cause AutoNom to abandon naming. The most relevant messages are the following:

- Isotope(s) in structure.
- Radical(s) in structure.
- More than two components in structure.
- Too many atoms.
- Triple bonds in a ring system with partial hydrogenation.
- Too many substituents on a single ring: currently up to 44 allowed.
- Too many atoms in a ring system: currently up to 44 allowed.
- Too many atoms in a ring assembly: currently up to 44 allowed.
- Too many atoms in a chain: currently up to 44 allowed.
- Branching ring assembly: currently only linear assemblies allowed.

- Ring assembly contains too many rings: currently up to 8 allowed.
- Cumulative double bonds in a ring system.
- Name too long.
- Inorganic structures.

Most of these limitations are only of the technical type (sizes of rings and ring assemblies, the numbers of substituents on rings or chains, etc.) and were introduced after careful statistical analysis of the material in the Beilstein database. They are a compromise between the hardware resources of an average personal computer for which AutoNom was designed, the statistical irrelevance of the classes of compounds in which they occur, and the expected success rate of the program.

Summary

The AutoNom program has more than 51,000 lines of code distributed in more than 460 various routines and functions. As commercial software, it is distributed in Europe by Beilstein Information Systems GmbH (Frankfurt/Main, Germany) and in the United States by Beilstein Information Systems, Inc. (Englewood, CO). AutoNom, as a so-called "draw and name" package, is available in a PC Windows version and a Macintosh version. As a batch version, it is available for PC as well as for VAX-VMS and ALPHA-OpenVMS minicomputers.

References

1. Davis, C. H.; Rush, J. E. *Information Retrieval and Documentation in Chemistry;* Greenwood Press: London, 1988; p 143.
2. Weininger, D. *J. Chem. Inf. Comput. Sci.* **1988**, 28, pp 31–36.
3. Barnard, J. M.; Jochum, C. J.; Welford, S. M. "Universal Structure/Substructure Representation for PC-Host Communication." In *Chemical Structure Information Systems. Interfaces, Communications, and Standards;* Warr, W. A., Ed.; ACS Symposium Series 400; American Chemical Society: Washington, DC, 1988; pp 76–81.
4. Goodson, A. L. *J. Chem. Inf. Comput. Sci.* **1980**, 20, p 167.
5. Proceedings of the Second International Conference, Noordwijkerhout, Netherlands, 3–7 June 1990. In *Chemical Structures 2. The International Language of Chemistry;* Warr, W. A., Ed.; Springer-Verlag, Berlin, Heidelberg, 1993.
6. "Chemical Nomenclature into the Next Millenium—Has It a Role?" In *Conf. Chemical Nomenclature, London, November 1987;* Laboratory of the Government Chemist: London.
7. Silk, J. A. *J. Chem. Inf. Comput. Sci.* **1981**, 21, pp 146–148.
8. Garfield, E. *J. Chem. Doc.* **1962**, 7, pp 177–179.
9. Garfield, E. *Nature* **1961**, 192(4798), pp 192–194.
10. Mockus, J.; Isenberg, A. C.; Van der Stouw, G. G. *J. Chem. Inf. Comput. Sci.* **1981**, 21, pp 183–195.

11. Smith, D. A. *J. Am. Chem. Soc.* **1992**, *114*(26), pp 10680–10681.
12. Glennon, R. A. *J. Med. Chem.* **1992**, *35*, pp 4918–4922.
13. Kroos, R. *CLB Chemie in Labor und Biotechnik*, **1992**, *8*, pp 466–467.
14. Lukomski, P. *Kurier Chemiczny*, **1991**, *6*, pp 16–18.
15. Kernytsky, B. S.; Mull, K. B. *CINF Abstract #88*. 206th American Chemical Society National Meeting, Chicago, IL, 1993.
16. Pouchert, C. J., Vice President, Aldrich Chemical Company, Inc., private communication, June 1994.
17. Wisniewski, J. L. *J. Chem. Inf. Comput. Sci.* **1990**, *30*, pp 324–332.
18. Wisniewski, J. L.; Goebels, L.; Lawson, A. In *Software Development in Chemistry 4;* Gasteiger, J., Ed.; Springer-Verlag: Berlin, Heidelberg, 1990; pp 19–29.
19. Wisniewski, J. L. *Beilstein Brief*, **1991**, *1*, p 2.
20. Goebels, L.; Lawson, A. J.; Wisniewski, J. L. *J. Chem. Inf. Comput. Sci.* **1991**, *31*, pp 216–225.
21. Wisniewski, J. L. CHEMTECH, **1993**, *23*, pp 14–16.
22. Wisniewski, J. L. In *Chemical Structures 2*, Warr, W. A., Ed.; Springer-Verlag: Berlin, Heidelberg, 1993, pp 55–63.
23. Wisniewski, J. L. In *Recent Advances in Chemical Information II*, Collier, H., Ed.; Royal Society of Chemistry, Cambridge, 1993; pp 77–87.
24. *Nomenclature of Organic Chemistry, Sections A, B, C, D, E, F, and H, 1979 ed.*, International Union of Pure and Applied Chemistry, Commission on the Nomenclature of Organic Chemistry: Pergamon Press, Oxford, 1979.
25. Lozac'h, N.; Goodson, A. L.; Powell, W. H. *Int. Ed. Eng.* **1979**, *18*, pp 887–899.
26. Hirayama, K. *The HIRN System Nomenclature of Organic Chemistry;* Maruzen: Tokyo; Springer-Verlag: Berlin, 1984.
27. *IUPAC Nomenclature of Organic Chemistry, Sections A and B, 1st ed.;* Butterworths: London, 1958.
28. *A Guide to IUPAC Nomenclature of Organic Compounds, Recommendations 1993;* International Union of Pure and Applied Chemistry, Organic Chemistry Division, Commission on the Nomenclature of Organic Chemistry (II.1); Blackwell Scientific Publications: Oxford, 1993.
29. Jochum, C. *CINF Abstract #28*. 192nd American Chemical Society National Meeting, Anaheim, CA, 1986.
30. Cooke-Fox, D. I.; Kirby, G. H.; Rayner, J. D. *J. Chem. Inf. Comput. Sci.* **1989**, *29*, 101–105.
31. Cooke-Fox, D. I.; Kirby, G. H.; Rayner, J. D. *J. Chem. Inf. Comput. Sci.* **1989**, *29*, pp 112–118.
32. Cooke-Fox, D. I.; Kirby, G. H.; Rayner, J. D. *J. Chem. Inf. Comput. Sci.* **1990**, *30*, pp 122–127.
33. Cooke-Fox, D. I.; Kirby, G. H.; Rayner, J. D. *J. Chem. Inf. Comput. Sci.* **1990**, *30*, pp 128–132.
34. Conrow, K. *J. Chem. Doc.* **1966**, *6*, pp 206–212.
35. Van Binnedyk, D.; Mackay, A. C. *Can. J. Chem.* **1973**, *51*, pp 718–723.
36. Rücker, G.; Rücker, C. *CHIMIA* **1990**, *5*, pp 116–129.
37. Babic, D.; Balaban, A. T.; Klein, D. J. *J. Chem. Inf. Comput. Sci.* **1995**, *35*, pp 1515–1526.
38. Babic, D.; Trinajstic, N. *Fullerene Sci. Technol.* **1994**, *2*, pp 343–356.
39. Eckroth, D. R. *J. Chem. Educ.* **1993**, *70*, pp 609–611.

40. Röse, P., Technische München Universität, *private communication*, February 1990.
41. Davidson, S. *J. Chem. Inf. Comput. Sci.* **1989**, *29*, pp 151–155.
42. Meyer, D. E.; Gould, S. R. *Am. Lab.* **1988**, *20*(11), pp 92–96.
43. *ISI Chemical Structure Dictionary*; ISI: Philadelphia, 1974.
44. Raymond, K. W. *J. Chem. Inf. Comput. Sci.* **1991**, *31*, pp 270–274.
45. Brockwell, J.; Werner, J.; Townsend, S. *Beaker—An Expert System for the Organic Chemistry Student*; Brooks/Cole: Pacific Grove, CA, 1989.
46. Navech, J.; Despax C. *New J. Chem.* **1992**, *16*, pp 1071–1076.
47. Dalby, A.; Nourse, J. G.; Hounshell, W. D.; Gushurst, A. K. I.; Grier, D. L.; Leland, B. A.; Laufer, J. *J. Chem. Inf. Comput. Sci.* **1992**, *32*, pp 244–255.
48. Gray, N. A. B. *Computer-assisted Structure Elucidation*; Wiley Interscience: New York, 1986.
49. Downs, G. "Rings—The Importance of Being Perceived." Proceedings of the Second International Conference, Noordwijkerhout, Netherlands, 3–7 June 1990. In *Chemical Structures 2: The International Language of Chemistry*; Warr, W. A., Ed.; Springer-Verlag: Berlin, Heidelberg, 1993; pp 207–219.
50. Gasteiger, J.; Ihlenfeldt, W. D. "Similarity Criteria for Chemical Structures and Reactions." Proceedings of the Second International Conference, Noordwijkerhout, Netherlands, 3–7 June 1990. In *Chemical Structures 2. The International Language of Chemistry*; Warr, W. A., Ed.; Springer-Verlag: Berlin, Heidelberg, 1993; pp 423–438.
51. Barth, A.; Westermann, U.; Pasucha, B. *J. Chem. Inf. Comput. Sci.* **1994**, *34*, pp 29–38.
52. Wisniewski, J. L. *J. Inf. Sci.* **1987**, *13*, pp 159–164.
53. Willet, P. *J. Chemom.* **1987**, *1*, pp 139–155.
54. Tannenbaum, A. M.; Augenstein, M. J. *Data Structures Using Pascal*; Prentice-Hall, Englewood Cliffs, 1981; pp 252 and 318.
55. Smith, P. A. S. *J. Chem. Educ.* **1992**, *69*, pp 877–878.
56. "Minutes of CNOC Meeting in Lisbon, Portugal, August 4–8, 1993"; Commission on the Nomenclature of Organic Chemistry, International Union of Pure and Applied Chemistry.

Index

A

Academia, Beilstein introduction, 136–139
Adjacency condition, substructure searching, 55
Atom-by-atom connectivity search, functional group recognition in AutoNom, 178–179
Atom mapping, CrossFire*plus*Reactions database, 106–107, 113–114
Atomic number vector (AN_v), input information for AutoNom, 172–174, 178
AutoNom (Automatic Nomenclature)
algorithm performance evaluation, 193–197
atoms allowed in organic compounds, 196*t*
automatic name construction from structural diagrams, 171–175
complexity of symmetrical ring systems, 181–183
diagnostic messages for rejected structures, 194–197
functional group recognition, 178–180
handling of tautomers, 173*f*
hash coding to shorten search times, 178
industrial evaluation by Wyeth Ayerst Research, 167
name assembly phase, 188–192
name tree creation, 186–188
parent structure selection, 184–186
program distribution, 197
ring system perception phase, 175–178
ring system recognition, 180–184
schematic of algorithmic naming phases, 174*f*
Sigma–Aldrich Chemical Company naming plans, 167

Smallest Set of Smallest Rings (SSSR) concept, 176*f*
software overview, 5, 11
structure-to-name translator by Beilstein, 166–167
substitutive nomenclature implementation, 192–193
user interface of program, 172*f*
See also Chemical nomenclature

B

Back references, *Beilstein Handbook of Organic Chemistry*, 18
Backtracking search technique
ChemFinder by CambridgeSoft, 62
operational systems, 59
RS3 Discovery by Oxford Molecular, 62
substructure searching, 57–59
Beilstein Brief, industrial experiences with CrossFire, 150
Beilstein Commander (CrossFire)
context controller, 80–81
hyperlink technology, 140
Beilstein CrossFire
academic access before CrossFire, 134–136
benefit of Committee for Institutional Cooperation (CIC) contract, 145
computerization impact, 135
industrial integration challenges, 154–156
operational experience at ETH Zürich, 124–126
reception and use in academia, 136–139
renaissance with CrossFire, 133–134
response to information explosion, 134–136
See also CrossFire applications, CrossFire structure search engine, CrossFire system

201